The Cathedrals of Pisa, Siena and Florence

Built Heritage and Geotechnics
Series editor: Renato Lancellotta

The Restoration of Ghirlandina Tower in Modena and the Assessment of Soil-Structure Interaction by Means of Dynamic Identification Techniques
Rosella Cadignani, Renato Lancellotta and Donato Sabia

Correction of Differential Settlements in Mexico City's Metropolitan Cathedral and Sagrario Church
Efraín Ovando-Shelley and Enrique Santoyo

The Tower of Pisa
History, Construction and Geotechnical Stabilization
J.B. Burland, M.B. Jamiolkowski, N. Squeglia and C. Viggiani

The Cathedrals of Pisa, Siena and Florence
A Thorough Inspection of the Medieval Construction Techniques
Pietro Matracchi and Luca Giorgi

For more information about this series, please visit: www.routledge.com/Built-Heritage-and-Geotechnics/book-series/BHG

The Cathedrals of Pisa, Siena and Florence

A Thorough Inspection of the Medieval Construction Techniques

Pietro Matracchi
Department of Architecture, University of Florence, Italy

Luca Giorgi
Department of Architecture, University of Florence, Italy

With the contribution of
Silvia Piccirilli

CRC Press
Taylor & Francis Group
Boca Raton London New York

CRC Press is an imprint of the
Taylor & Francis Group, an **informa** business

CRC Press/Balkema is an imprint of the Taylor & Francis Group, an informa business

© 2022 Taylor & Francis Group, London, UK

Typeset by codeMantra

Library of Congress Cataloging-in-Publication Data
Names: Matracchi, Pietro, author. | Giorgi, Luca, author. | Piccirilli, Silvia.
Title: The cathedrals of Pisa, Siena and Florence : a thorough inspection of the medieval construction techniques / Pietro Matracchi, Department of Architecture, University of Florence, Italy; Luca Giorgi, Department of Architecture, University of Florence, Italy ; with the contribution of Silvia Piccirilli.
Description: Leiden : CRC Press/Balkema, [2021] | Series: Built heritage and geotechnics, 2640-026X ; volume 4 | Includes bibliographical references and index.
Subjects: LCSH: Duomo (Pisa, Italy) | Santa Maria del Fiore (Cathedral : Florence, Italy) | Duomo di Siena. | Building—Italy—History—To 1500.
| Civil engineering—Italy--History—To 1500.
Classification: LCC NA5621.P713 M38 2021 (print) | LCC NA5621.P713 (ebook) | DDC 726.60945--dc23
LC record available at https://lccn.loc.gov/2020056836
LC ebook record available at https://lccn.loc.gov/2020056837

Published by: CRC Press/Balkema
              Schipholweg 107C, 2316 XC Leiden, The Netherlands
              e-mail: Pub.NL@taylorandfrancis.com
              www.crcpress.com – www.taylorandfrancis.com

ISBN: 978-0-367-17487-3 (Hbk)
ISBN: 978-1-032-01920-8 (Pbk)
ISBN: 978-0-429-05704-5 (eBook)

DOI: 10.1201/9780429057045
DOI: https://doi.org/10.1201/9780429057045

Contents

Series editor foreword

This is the fourth of a series of volumes on Built Heritage and Geotechnics, intended to reach a wide audience: professionals and academics in the fields of civil engineering, architecture and cultural resources management, and particularly those involved in art history, history of architecture, geotechnical engineering, structural engineering, archaeology, restoration and cultural heritage management, and even the wider public.

Motivations of this series rely on the fact that preservation of built heritage is one of the most challenging problems facing modern civilization. It involves in inextricable patterns various cultural, humanistic, social, technical and economic aspects. The complexity of the topic is such that a shared framework of reference is still lacking among art historians, architects, and structural and geotechnical engineers. This is proved by the fact that although there are exemplary cases of an integral saving of any structural components with its static and architectural function, as a material witness of the knowledge, the culture and the construction techniques of the original historical period, there are still examples of uncritical confidence in modern technology which leads to the replacement of previous structures with new ones, which only preserve an iconic appearance of the original monument.

For these reasons, publishing short books on specialized topics, like well-documented case studies of restoration works at one specific site or of a monument, or detailed overviews of construction techniques, intended as a material witness of knowledge of the historical period in which the monuments were built, may be of great value.

In this context, the present volume, authored by Pietro Matracchi, Luca Giorgi and Silvia Piccinini, is a thorough inspection of the medieval construction techniques offered to readers by considering three examples: the cathedrals of Pisa, Siena and Florence.

It is a matter of really unique and capturing examples in many respects, also because few buildings reveal so much of a city history and civic aspirations as these cathedrals.

The construction of the cathedral of Pisa, which is a component of the Pisa amazing religious core, the so-called "Campo dei Miracoli", that should be better addressed to as "Campo delle Mirabilia", being a sight whose impact no amount of prior knowledge can blunt, dates from the period of Pisa greatest prosperity and power, from the 11th to the 13th centuries: the cathedral was begun in 1063 and completed at the end of the 12th century; the baptistery was started in 1152; the bell tower was added in 1173 and the cemetery, the "Camposanto", at the end of the 13th century. The Pisan Romanesque architecture of this period, distinguished by its white-and-black marble facades, is complemented by the impressive and finest medieval sculpture from the workshops of Nicola and Giovanni Pisano.

The cathedral of Siena has an impressive and without parallel history, because of the challenging problems that the builders had to face with the ongoing construction. The authors of this book clearly highlight how, because of the complexity of the genesis of this cathedral, we cannot understand the construction based on a project as we would understand today but is rather the achievement of many builders that were able to find solutions, learning by going and by taking into account the construction site constraints. There was a previous cathedral, built in the 11th century, whose location was dictated by the underlying soil profile: a layer of loose filling material overlying a conglomerate divided into two parts by a fault line, the western side lying about 5 m higher than the eastern portion. This proves why the original cathedral, as testified by the remains of its apse, was much shorter than the present-day cathedral, being located on the upper part of conglomerate; on the contrary, the builders laid the foundations of the crypt of the new cathedral and the Baptistery that today we observe, below this level.

The construction of the new cathedral can presumably be dated in the period 1120–1260 and was subjected to constant plans of expansion throughout Siena's years of medieval prosperity.

In particular, an extraordinary change was made at the beginning of the 14th century, because on 23 August 1339, there was an extremely ambitious deliberation to attempt to double its extent by building the so-called "Duomo Nuovo", a re-oriented new cathedral, orthogonally to the existing one, by using its nave as a transept and building a new nave out towards the "Campo".

At this point, the reader's attention will certainly be captured by the description of the technical difficulties that the builders had to be faced, related to the connection between the Duomo Nuovo and the earlier cathedral in the crossing area. Masters Domenico di Agostino and Niccolò di Cecco del Mercia were asked for an opinion in 1356, the debate about the suggested solutions being referred in detail in Chapter 3.

Worth noting, in 1357, while in Florence, the building of a basilica body, whose dimensions had no comparison, was in progress, the Florentines Francesco Talenti and Benci di Cione were appointed in Siena to give an assessment about the stability of the "Duomo Nuovo", in danger of collapse. They claimed to demolish four columns, with the arches and vaults resting on them, and suggested dismantling nine bays.

The structures that today we observe are the remains of a basilica body, at that time almost completely built, after collapses and the dismantling of the unsafe parts.

And it is an amazing story for the reader to discover that just few weeks later, Francesco Talenti and Benci di Cione used the experience gained in Siena to solve the problems posed by the building of the naves of the cathedral in Florence.

The decision to build the present-day Cathedral of Santa Maria del Fiore in Florence was reached through a gradual process, preceded by attempts to renovate the existing Cathedral of Santa Reparata, founded in the 6th century. At the end of the 13th century, it appeared inadequate, both in comparison to the imposing nearby Baptistery, and also in comparison to the new churches being constructed by the mendicant orders, such as the Dominican Santa Maria Novella and the Franciscan Santa Croce.

After a first attempt to work on renovation of Santa Reparata around 1293–1296, it was decided to build a new cathedral, partially superimposed on the old structure, setting back the façade, but greatly expanded to reach its current size.

The work progressed fairly slowly until the sixth decade of the 14th century. As already mentioned, in 1357, Francesco Talenti and Benci di Cione were asked for an assessment of the Duomo Nuovo in Siena, a unique opportunity to study errors made in the construction of a great cathedral at its real scale. What a physical model, at a 1:1 full scale! It was the knowledge acquired in that occasion by the two Florentine masters that allowed to start and complete within a few years the large square bays, characterized by pillars span of almost 20 m, representing an authentic construction primacy.

By 1418, the nave was finished, the tribunes were completed, and a drum was in place to bear the weight of Brunelleschi dome, whose conception was magnificent: it rises from a base some 55 m above the floor of the nave and spans a distance of 45 m. On 25 March 1436, the Annunciation Day, its completion was marked by the consecration of the cathedral.

It is apparent from these few remarks the challenging problems that architects and mesons had to face, and till today, their work and their skill in finding solutions, mostly innovative, are not even envisaged.

Therefore, the reader will discover this book as an invaluable guide, because the construction techniques and concepts of the cathedrals of Pisa, Siena and Florence are examined in detail, based on new data and using a methodological architectural diagnostics approach.

New detailed surveys, carried out using advanced tools, together with direct and in-depth inspections to examine all parts of the buildings, have enabled the authors to identify the building phases and the different construction techniques used each time. The information thereby acquired also formed the basis for a new interpretation of the archival documents.

Accordingly, the major problems encountered and the solutions adopted in the three cathedrals are properly addressed: in Pisa, the construction of the elliptical dome above the rectangle consisting of six thin pillars; in Siena, the design changes from the first system in the 13th century to the "Duomo Nuovo", and the structural adaptations and devises following earthquakes (particularly after the earthquake of 26 May 1798), that medieval builders had chosen not to use; and the specific construction solutions adopted in Florence to avoid the instability problems that could have occurred in the construction of the large bays of the basilica body.

The book originates from a long and well-established research activity carried out by the authors, and their great experience in the field is reflected in the clarity of the text and the richness of illustrations.

The adopted methodological approach has led to an understanding of the actual structure of the three cathedrals, an essential basis for a correct evaluation of the state of conservation of materials and structures for any restoration work.

In short, this volume cannot be missed by people involved in art history, architecture, engineering, and restoration work, and it will also attract the attention of a wider audience interested in gaining a deeper understanding of these "mirabilia".

The series editor Renato Lancellotta

Acknowledgements

It is almost impossible to mention all the people who have supported this research work, carried out over several years, without forgetting someone. In any case, we would like to mention the Opera della Primaziale Pisana, the Opera della Metropolitana in Siena and the Opera di Santa Maria del Fiore in Florence, and to thank their presidents, management bodies, members of the technical offices and caretakers. We are grateful to the staff of the Surveying Laboratory of the Department of Architecture of Florence University, in particular Giovanni Anzani and Francesco Tioli who carried out the laser scanner survey, respectively, in the transept of Pisa cathedral and the aisles of Florence cathedral. We would like to thank Sara Marianelli for drawing the intrados profiles of the vaults in the aisles of Santa Maria del Fiore based on the 3D model, and Elisa Targetti for developing a section from the laser scanner survey of Giotto's Bell Tower. A big thank you goes to Adriano Bartolozzi for providing some photos of Pisa and Siena cathedrals which he took when he was the head of the Photographic Laboratory of the Institute of Restoration of Florence University.

About the authors

Pietro Matracchi is an architect and Associate Professor at the Architecture Department of the University of Florence, where he has taught Architectural Restoration since 2002. He taught the same subject at the Universities of Pisa and Perugia. He is the author of papers and monographs on architectural diagnostics aimed at the study of construction techniques, construction phases, and conservation problems of materials and structures. The subjects he has covered include the Museo Nazionale del Bargello in Florence, the cathedrals of Perugia, the Baptistery and Palazzo Pitti in Florence, the Medici villa in Poggio a Caiano, Santa Maria della Consolazione in Todi, theatre architecture of the 19th century, and architectural works by Francesco di Giorgio Martini and Giorgio Vasari. In his research, he has paid special attention to the philological problems in the interpretation of surveys and concrete architectural artefacts, in relation to archival documents.

He is the co-author, with Luca Giorgi, of studies on architectural complexes extended on urban scale as the towers of San Gimignano.

Luca Giorgi is an architect and former Associate Professor of Architectural Restoration at the University of Florence. He investigated cultural heritage using NDT techniques, performing IR-reflectography on paintings (Michelangelo's Tondo Doni and Botticelli's Birth of Venus), thermography on buildings (S. Maria del Fiore, Spedale degli Innocenti, Palazzo dei Cavalieri in Pisa), and art objects (bronzes in the National Museum in Florence and the Paradise Gate of the Florence Baptistery). He is the author of publications based on direct assessments of some of the most relevant Italian civil (Strozzi and Bargello Palaces in Florence) and religious buildings, such as the complex of S. Maria del Fiore in Florence, S. Lorenzo Maggiore in Milan, S. Pietro in Rome, S. Biagio in Montepulciano, Jesuit churches (Gesù in Montepulciano, San Rocco in Parma), and the Cathedral of Siena. He was also promoter, national delegate and vice-chairman of the COST-UCE Group "Urban Heritage – Building maintenance".

Chapter 1

Introduction

Medieval cathedrals have always sought to represent the maximum architectural expression of the cities in which they stood, all the while seeking to rival similar buildings in other places.

The extraordinary characteristics of cathedrals in terms of the richness of materials and the drive to come up with innovative architectural solutions made for an extremely complex construction, which generally lasted for centuries.

While they were being built, there was no supposed original design to follow which would have provided all the necessary information; rather, the key decisions on the choice and production of the materials and the constructive solutions were made at the construction site itself. Furthermore, decisions about the architectural structure to be created were often made step-by-step during the execution phases, often demonstrating a willingness to make changes that were capable of calling previous choices into question. The construction site was a place where architectural experimentation occurred.

All these aspects emerge strongly from the studies conducted on the cathedrals of Pisa, Siena and Florence presented in this book. The research results were obtained through architectural diagnostics based on a direct accurate analysis of the structures and an interpretation of the surveys, conducted with traditional techniques and laser scanners. In this latter case, the data provided by 3D modelling was processed to obtain the diagrams, sections and elevations deemed necessary to understand the real, authentic structure of these churches. The architectural diagnostics carried out led to the observation of specific aspects of these buildings without losing sight of their general architectural context, and when the general structure was taken into consideration, we also contemplated the multiple specificities linked to an extremely complex building structure, where countless contributions have become stratified over the centuries.

Whenever possible, the information was supplemented with the results of non-destructive or moderately invasive instrumental investigations, such as thermographic imaging, dendrochronology, the laboratory analyses of mortars, ground-penetrating radar, flat-jack tests, core drilling, measuring the tensional states of metal ties and information on the structure of the subsoil. Other data were obtained from structural modelling, available for the dome of Siena cathedral, and for the transversal openings of the basilican body of Santa Maria del Fiore.

The studies conducted an in-depth analysis of the complex construction genesis of the cathedrals, considering the operational conditions faced by the builders at these construction sites. The information obtained from a similar approach was consulted in relation to archive documents, in many cases giving rise to new interpretations more pertinent to what actually occurred at the construction sites and the works they refer to.

This research based on architectural diagnostics had two main purposes. One was to learn more about the construction history and transformations of the cathedrals of Pisa, Siena and Florence. The aim was to cast aside idealizations by paying constant attention to the concrete elements – the materials, construction solutions and state of conservation – making up these buildings of extraordinary cultural value. The genesis of these churches has a multifaceted character and involved many builders and the development of several designs that accompanied the work. The other fundamental goal was to provide appropriate knowledge in order to understand and address the conservation problems of such complex buildings, above all regarding the constructive structural aspects. Each of these churches, beyond their apparent homogeneity, is in fact characterized by many construction aspects, which have sometimes changed significantly over time, knowledge of which becomes vitally important when it comes to assessing any structural vulnerabilities. It should be pointed out that the vulnerabilities are rarely the result of a systemic condition and are more frequently linked to specific local situations, caused by particular episodes. Precisely due to their specific character, in order to identify local vulnerabilities, an accurate diagnostic study commensurate with the specific characteristics of the building is required.

This book is divided into three chapters, each dedicated to one cathedral. The areas covered in each chapter follow the course of events and are placed in chronological order.

In Pisa cathedral, the transept was studied, and in particular, the complex system of construction solutions used in the transition from the rectangular crossing to the elliptical plan of the dome above. In the drum area, a surprising variety of construction systems were identified, which allowed the builders to gradually create a connection between the crossing and the dome. In addition to the pendentives, a set of overlapping arches were created, in some cases skewed and partially overhanging; walls were built with variable inclinations, and in some cases, the archivolts of the arches were also raised following inclined planes. The complex relationship between form and structure was confirmed by the detailed analysis of the masonry structure of the pendentives, documented for the first time with extremely accurate surveys, the forerunner of which was Piero Sanpaolesi (1975). Among the apparently similar pendentives, significant differences were detected in the masonry, reflecting the gradual definition of the construction solutions developed at the site during the execution phases. The stability of the crossing is also ensured by the use of flying buttresses with two archways, placed next to the transverse arches of the crossing, the existence of which in the false ceilings of the matronea was noted for the first time by Adriano Peroni (1995, 49–50).

Siena cathedral is perhaps a unique example for the continuous innovative drive to alter projects that were in progress or architectural arrangements that had already been completed. On these events, a long and in-depth study campaign promoted by Peter Anselm Riedl and Max Seidel, published in several volumes on the surveys (Haas & von Winterfeld, 1999b), the architectural analysis (Haas & von Winterfeld, 2006a, 2006b), and the archive documents (Giorgi & Moscadelli, 2005), marked a turning point. The topic was resumed with new surveys and studies, placing particular attention on the medieval construction sites that led to the construction of the intermediate cathedral in the sixth decade of the 13th century, and the events of the "New Cathedral" which concluded with the unfortunate outcome of its collapse due to the inadequacy of the construction solutions put in place.

The Santa Maria del Fiore complex – made up of the baptistery, cathedral and bell tower – was the subject of an extensive and complex programme of studies and surveys promoted by Giuseppe Rocchi Coopmans de Yoldi, published in four folio volumes (Rocchi et al., 1988; Rocchi Coopmans de Yoldi, 1996, 2004, 2006) containing contributions from the authors of this publication. In this volume, attention has been placed on the cathedral construction site, starting with the renewal of the ancient cathedral of Santa Reparata at the end of the 13th century, followed by the construction of part of the façade and the flanks of the present-day cathedral and a subsequent period when work came to a standstill, during which most effort was concentrated on the construction of the bell tower, which was then completed with the contribution of Giotto, Andrea Pisano and Francesco Talenti. As regards the cathedral, 1357 was a year of great changes which led to the creation of a vaulted basilican body with a central nave with four bays, with an average centre-to-centre distance between the pillars of 19.60 m. The exceptional nature of this construction dimension required, in addition to the use of flying buttresses hidden in the garrets of the side aisles, the preparation of temporary works that guaranteed the stability of the structures during construction. So the central nave was fitted with temporary ties, the last of which were removed in Brunelleschi's time, while other similar ties with a temporary function, placed in the garret areas, were not removed.

In large buildings with vaulted systems, the problem of stability during construction was cogent in that very different static conditions were gradually obtained with respect to the final constructive structure. This is a highly interesting aspect of the construction work, but there are still few studies on it. Temporary ties were also placed in the central nave of the cathedral of Majorca (Roca et al., 2019, 160–161), which has rectangular bays but of a similar width (19.40 m) to those of Santa Maria del Fiore.

Interaction between the constructed parts and the parts under construction had significant consequences in the layout of Santa Maria del Fiore. When the construction of the dome reached the second corridor, that is two-thirds of its vertical height, there was subsidence in the pillars of the dome adjacent to the basilica body. This resulted in instability in the fourth bay of the central nave, traces of which can still be seen today. It was Brunelleschi himself who came up with the idea of the present-day double ties crossing the central nave and that, with wooden elements, continue into the garrets of the side aisles anchoring to the pillars of the flying buttresses. Brunelleschi also made changes to the flying buttresses between the second and third bays, adding small arches in an intermediate position to the pillars, thereby creating double level flying buttresses. This intervention was a big change with respect to the work of the medieval builders, who had wanted to construct the central nave without ties.

Much more incisive alterations were made on several occasions to the structure of the Siena cathedral, which was completed in the 14th century. In this case, it was the vulnerability of the structures brought to light by the numerous earthquakes that led, from the 16th to the 19th centuries, to a gradual but unstoppable conceptual rethinking of the entire constructive structure, which saw ties, buttresses and flying buttresses placed in the aisles. These were all constructive protections that the medieval builders had chosen not to use.

Another issue examined is the correlation between construction phases. In 1357, Francesco Talenti and Benci di Cione were called to Siena as consultants to give an opinion on the critical conditions of the "New Cathedral," which then collapsed a few months later. The two builders must have learned a great deal from this experience. At the same time, in Florence, it was decided to finish the construction of the bell tower, interrupting an additional level in order to begin building a vaulted basilican body considerably larger than that of the "New Cathedral." Benci di Cione, and Francesco Talenti in particular, played a leading role in this new start to the Florentine cathedral.

Chapter 2

Pisa cathedral

2.1 From the rectangular base of the crossing to the elliptical springer of the dome: architecture, building site and constructive solutions

As is the case for many buildings from the same era, documentary evidence on the origins and construction of Pisa cathedral was often incomplete or entirely missing. The 5-aisle basilican body, which extends beyond the crossing up to the semi-circular apse, and the arms of the 3-aisle transept joining the space surmounted by the dome are thought to be the result of complex building work started in the sixth decade of the 11th century and completed in the 1380s (Peroni, 1995, 14–15) (Figures 2.1 and 2.2).

The epigraphic evidence found in the cathedral is particularly important, recognizing the key role Buscheto played in the preparation of the project. He was succeeded by the master Rainaldo and thereafter by other builders. The cathedral was consecrated and was the seat of the Council in the presence of Innocent II in the first half of the 13th century (Ronzani, 1996, 35). The existence of the southern transept was also confirmed in this same period. The bronze door of the façade was built in 1181, which would seem to imply the completion of this part of the building. However, archaeological evidence was found that suggested there had been an extension to the west of the aisles and therefore that the basilican body was originally smaller (Bacci, 1917, 2; Scalia, 1982, 817–859; Milone, 2008, 73–74, 78). Work continued on the dome until the 1380s (Tanfani Centofanti, 1897, 439–442) (Figure 2.3).

In addition to the complexities of the building phases, transformations occurred which over the years have changed the structure of the church several times. An extremely important one was the fire in 1595, which destroyed a large part of the building and resulted in the consequent restoration of the wooden parts of the roof, as well as stone elements like the capitals (Peroni, 1995, 120).

The architecture of Pisa cathedral is an extraordinary complex subject. It is examined here through in-depth studies, in particular concerning the specific construction solutions of the domed space, considered in the inseparable relationship between architectural space, structure and the act of building (Figures 2.4 and 2.5).

The rectangular crossing is formed from the intersection between the five-nave basilica and the three-nave transept. The central nave of the longitudinal part of the building determines the width of the shorter side, while the longer side extends for the full width of the aisles of the transept (Figure 2.6).

On the four corners of the crossing, there are cruciform pillars, which on the shorter side support pointed arches that extend for the full width of the nave. These arches are of different

Figure 2.1 Pisa cathedral; view of the façade and north side.

Figure 2.2 Pisa cathedral; view of the transept.

Figure 2.3 Nave and apse of Pisa cathedral.

heights, and the one facing the basilican body is higher. On the longer side of the crossing, there are two monolithic granite columns placed in continuity with those of the nave of the transept. Those of the crossing, however, have a larger diameter, equivalent to that of the column of the central nave, measuring approximately 1 m (Figure 2.7).

The rectangular space within these bounds is regular, with almost equivalent opposite and diagonal sides. The new surveys instead highlight a unique characteristic of the transept arms: they have two orthogonal sides, the east sides with the respective end walls with apses and the western sides which become narrower as they approach the crossing gradually reducing the width of the adjacent aisles. In particular, the walls just mentioned are rotated

Figure 2.4 Laser scanner survey of the external area at the level of the roof of the aisles.

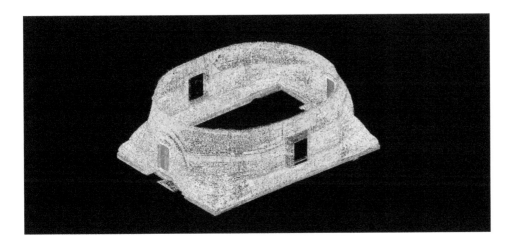

Figure 2.5 3D model of the laser scanner survey inside the drum.

1° in the north arm and 2° in the south arm. In the latter case, in the west aisle, the gradual narrowing in the first four bays is particularly marked, reaching a maximum of around 45 cm.

A similar alignment of the transept arms would suggest that the construction of the entire three-nave part of the building was not coeval and that the positioning of the western wall depended on the dimensions of the crossing. We wanted to emphasize this aspect to demonstrate how complex the construction of the cathedral may have been, even if the purpose of this work was not to identify the construction phases of the ground plan.

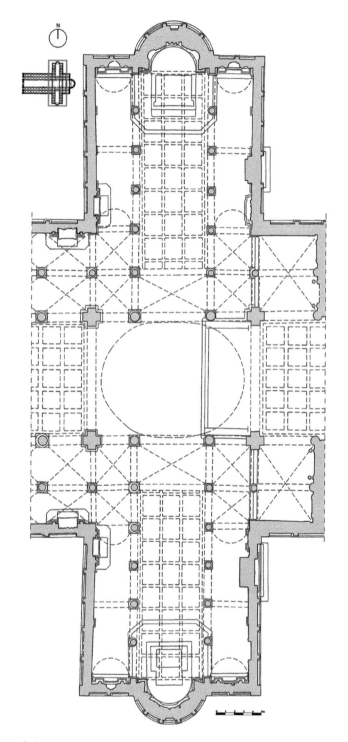

Figure 2.6 Plan of the transept and crossing.

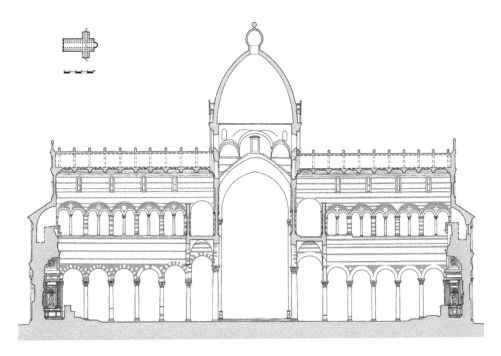

Figure 2.7 Longitudinal section of the transept.

The accentuated longitudinal development of the rectangle inside the crossing, roughly 12.50 m wide and 17.20 m long, stems from specific constructive relationships with the adjacent parts of the building. The cruciform corner pillars are aligned transversely with the sides of the transept and longitudinally with the arches corresponding to the central nave.

The intermediate columns on the longer sides of the crossing are aligned with those of the naves of the transept (Figure 2.8). The structural continuity of these elements is highlighted by the longitudinal section of the transept: the columns bear arches with a wall of considerable height above, interrupted at the level of the matroneum, which forms a continuous diaphragm arch with the sides of the adjacent central naves of the transept. The arches above, which frame the mullioned windows of the matronea, create a second system of diaphragm arches whose archways are of a more uniform height. This double level of diaphragm arches therefore sits just above the columns and above the pillars that continue to the level of the matronea (Figure 2.9).

At the level of the side aisles, the corner pillars are also flanked by transversal diaphragm arches which rise up to floor level of the matronea; in correspondence to the pillars, with respect to the columns, the springers of the diaphragm arches are lower to create a diaphragm wall of greater height. This creates constructive protections halfway up the crossing which help to make the columns and pillars of the crossing more stable. The corner pillars are also longitudinally flanked by a double system of diaphragm arches, one above the arcades of the central nave and the other higher up above the arches of the mullioned windows of the matronea (Figure 2.10). The latter is higher and rises beyond the pillars reaching the base of the drum. In other words, the intermediate area of the shaft of each of the corner pillars of the crossing, which are apparently very slender, is buttressed on three sides by diaphragm

Figure 2.8 Longitudinal section of the crossing.

arches. This means that the unbuttressed parts of the pillars are actually quite short, even if, when observed from the central nave, they appear as remarkably slender unbuttressed pillars from the base to the capital, supporting the transversal arches of the crossing.

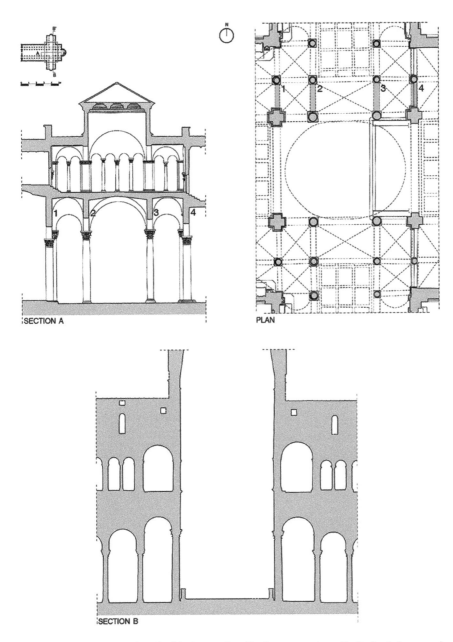

Figure 2.9 Section A and plan, highlighting the diaphragm arches (1, 2, 3, 4, in green), and section B.

On the longer sides of the crossing, at the level of the matronea, there is a quadrifora flanked by mullioned windows, which respectively correspond to the side aisles and the central nave of the transept. Higher up in the centre of the same walls, an arch was created. It was then infilled with masonry and has an archivolt opening, in turn infilled by a thin brick wall with a small door (Figure 2.11). This arch, which is not visible from the crossing

Figure 2.10 Longitudinal section of the structures flanking the crossing.

Figure 2.11 Detail of the south (a) and north (b) sides of the crossing from the level of the matronea to the drum. The following are indicated: the discharging arches (in green) at the level of the garrets of the naves of the transept with the archivolted doors (light green), the staircases on the east and west sides (pink) which lead to the garrets of the longitudinal body, the entrance doors (purple) from the external staircases positioned at the summit of the flying buttresses, the pendentives (yellow) with discharging arches above (brown), the discharging arch and the trifora at the level of the drum (both in light blue), the asymmetrical arches (red) above the pendentives.

where the walls are plastered and painted in bands of two colours, is slightly narrower than the one below the quadrifora, and the top part of it, with the archivolt, just extends beyond the height of the springer of the pendentives (Figure 2.7). It should be pointed out that the infill of the arches, made up of squared stone ashlars and visible from the transept garrets, was executed with the same material (a type of sandstone known as panchina livornese) and precision as the arches themselves. It should also be pointed out that the arch is about 20 cm thicker than the infill wall. All of this suggests that it was conceived as a unitary construction executed in different construction phases in order to obtain a discharging arch above the quadriforas, which in turn are above a further arch; the discharging arch is therefore positioned in an area probably thought to be structurally vulnerable (Figure 2.12).

The infill wall of the discharging arches has arched doors which open at a lower level than the springer of the same arches. This level is roughly equivalent to the start of the flights of stairs which, on the shorter sides of the crossing, rise as far as the garret of the central nave of the longitudinal body. It is plausible that between the lower part of the stairs and the passages in the infills of the discharging arches, there was a walkway connecting the levels of the trusses of the longitudinal and transverse central naves.

In several cases, for instance, in the churches of Santa Croce in Florence and San Francesco in Cortona, there is still evidence today of passageways made up of boards resting on the central area of the truss tie-beams so that maintenance work could be carried out on the roofs.

On the walls flanking the discharging arches, a further distinct construction solution should be highlighted. On the outer side of the walls, two arches of the blind loggia visible from the outside in the nave wall above the clerestory extend into the area of the dome. The columns of these loggias rest on wall offsets of approximately 30 cm, but this reduced thickness also affects an extensive part of the crossing wall. This construction aspect is important but barely noticeable as the only connecting elements between the interior and the exterior are the oculi, with different diameters, set within some small blind arches of these blind loggias (Figure 2.13).

Figure 2.12 Detail of the infilled wall of the discharging arch and the entrance to the garret of the nave of the north wing of the transept.

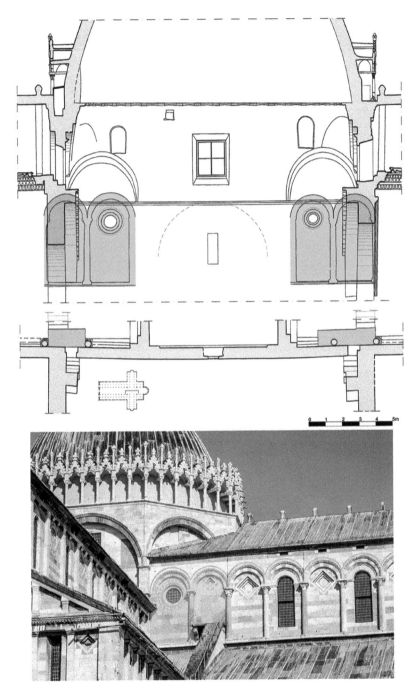

Figure 2.13 (a and c) Relationship between the crossing space and the blind arches of the clerestory (light brown) on the outside; (b) plan at the level of the clerestory with the landing (purple) between the external stair on the flying buttresses and the internal one above the transverse arches of the crossing.

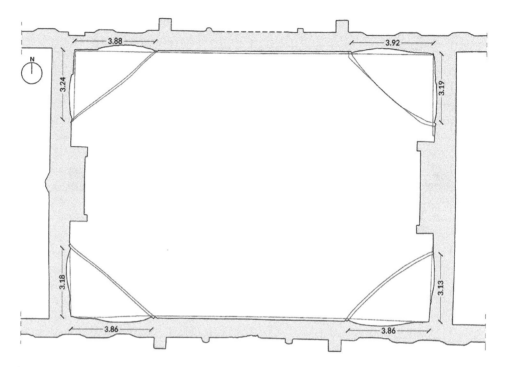

Figure 2.14 Plan at the base of the pendentives.

The horizontal cornice that separates the pendentives from the lower part of the crossing, indicating the start of the drum as an apparently independent element, marks an architectural delimitation that simplifies the real relationships between the interior and exterior construction and architectural elements. This level, in fact, is surpassed by both the top of the discharging arches of the north and south walls and by the arches of the external blind loggias. The latter partly overlap the masonry of the pendentives, which have a curved profile from the initial part of the plan. In the overlapping areas, the wall thicknesses are a minimum on average of approximately 55 cm, while the wall with no reductions is around 90 cm thick (Figure 2.14).

The constructive and architectural complexity is evident in the area where the longitudinal body and transept meet and in the crossing area where the base of the drum marks the start of the pendentives and the transition from the rectangular plan to the irregular octagonal plan occurs.

2.2 The flying buttresses

A further key construction element, belonging to the buttressing structures below the drum level, is the flying buttresses. The existence of this structural device was recently highlighted by Adriano Peroni, who identifies it as the archway close to the central nave, proposing an axonometric representation of it (Peroni, 1995, 80).

These are flying buttresses with sloping double arches, interposed between the large arches of the shorter sides of the crossing and the sides of the transept that act as buttresses.

Between the two archways of the flying buttresses is a column roughly corresponding to the column below interposed between the side aisles. It should be said that the column of the flying buttresses rests on a diaphragm arch of considerable height and extending to the two bays of the side aisles (Figure 2.15). As regards the continuity of the vertical structures, note that the crossing pillar is wider at the level of the matronea than it is at the lower level: the pillar is, respectively, approximately 2.60 and 2.00 m wide, but it should still be noted that these different sections of the pillar are interposed with diaphragm arches of considerable height (approximately 4.04/4.20/4.90/5.90 m). This means that the overhang of the pillars at the matronea level has no direct effect on the transverse arches of the aisles.

The flying buttresses abut the haunches of the transverse arches of the crossing; the masonry above them gradually increases in height and reaches the level of the blind loggias of the clerestory.

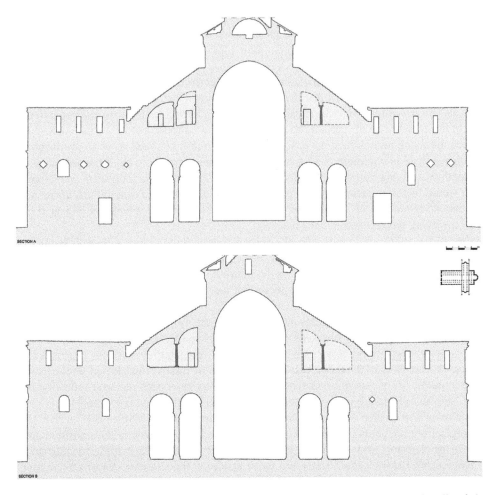

Figure 2.15 Longitudinal sections of the flying buttresses abutting the external walls of the transept.

The north-west flying buttress has been infilled by a thin wall which allows us to see the double arch on both sides, while the base and capital of the central column are only partly visible (Figures 2.16 and 2.17).

In the north-east flying buttress, the off-centre position of the infill leaves the central column, the outermost arch and the initial part of the internal one almost entirely visible, while the rest is covered up by coffering. A test carried out on the infill wall also revealed the corbel on which the outermost arch abutting the transept wall rests.

As regards the flying buttresses on the south side, the inspections were more complex as they have been completely infilled. There is only one trace in the south-west flying buttress, where a crack has formed in the plaster painted in light and dark bands near the intrados of the flying buttresses. Their presence was then confirmed by the inspection of the parts above the coffers, which in particular revealed the archway abutting the pillar of the crossing (Figure 2.18). The arcades of the flying buttresses and the pillars they abut are made of panchina livornese limestone, and the infill is brick. These materials are exposed above the coffering, while the visible parts of the matronea, with the extensive work carried out after the disastrous fire that ravaged the cathedral in 1595, were probably plastered and painted to mimic two colours of stone laid in horizontal black and white bands (Peroni, 1995, 44) (Figure 2.19), but these works were also carried out during the 19th-century restoration (Levi, 2015, 125–127; Calderoni Masetti, 2014, 40).

It was also possible to inspect the areas above the coffers that flank the south-east flying buttress, obtaining confirmation that the innermost archway had been infilled (Figure 2.18). It was also observed that the bearing ashlar of the flying buttresses is partly inserted into the pillar of the crossing; in the case of the south-east flying buttress, just below the connection between the flying buttress and the pillar, there is a thinner row made up of an overhanging moulding, placed in continuity with a row that continues along the wall of the central aisle. The south-west flying buttress abuts just above a course made from a different stone material; it would therefore seem that during the construction of the crossing pillars, a row was built that stood out for the material or wall structure characteristics and that indicated the height the flying buttress would have reached.

In the flying buttresses to the south, the existence of an intermediate support is borne out by the very curvature of the archways near the central nave and confirmed by a test carried out in the stone infill of the south-east flying buttress, which revealed the shaft of a fluted column. It is therefore reasonable to suppose that the column was interposed between two arches of the south-west flying buttress (Figure 2.20).

There are no documentary findings indicating when the flying buttresses might have been infilled. If anything, it can be observed that on the visible part of the capitals of the columns of the north-west and north-east flying buttresses, there are simplified scrolls, abacuses and rosettes. One archway of the flying buttress seems to show traces of blackened smoke marks on the intrados. All of this suggests that the capitals may have been replaced following the fire of 1595 and that the flying buttresses were also infilled after the same event.

A staircase was created in the masonry over the top of the flying buttresses (Figure 2.21). It emerges from the roofs and reaches the level of the blind loggias, in the transition area between the clerestory and the drum of the crossing. The cross section helps us to understand the relationship between these stairs and those inside on the shorter sides of the crossing (Figure 2.15). It is clear that the landing at the top of the flying buttresses was placed at the starting height of the stairs inside the crossing, which led to the garrets of the transept, most likely by means of a wooden walkway that no longer exists, and those of the central nave of the longitudinal part of the building (Figure 2.22 and 2.23).

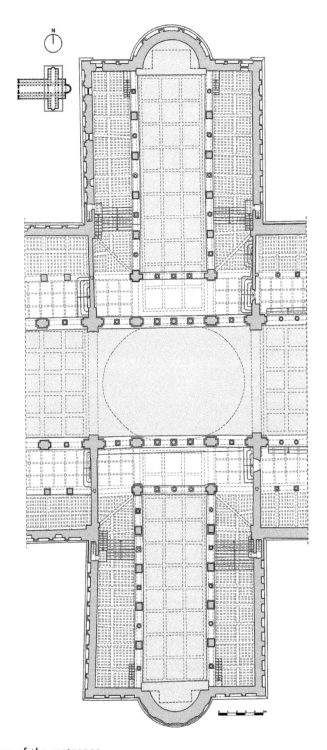

Figure 2.16 Plan of the matronea.

Figure 2.17 North-west flying buttress infilled at the matronea level.

Figure 2.18 (a) Connection between the south-east flying buttress and the crossing pillar. (b) Portion of the south-west flying buttress, infilled by a brick wall, visible in the matroneum roof space.

Figure 2.19 View of the matronea.

The flying buttresses, placed to counteract the walls above the transversal arches of the crossing, made it necessary to construct a landing that would create a direct connection with the overhanging stairs on the shorter sides of the crossing (Figure 2.24).

On the west side, the two doors providing access to the stairs of the crossing have jambs in keeping with the surrounding stone facing of large ashlars. The ashlars of the jambs are more often than not continuous with the rows of the adjacent masonry, and sometimes the ashlars continue from the jambs to behind the columns of the external arches of the clerestory. Jambs with independent ashlars are less frequent and are limited to partial segments (Figure 2.25).

The landings at the top of the stairs above the flying buttresses, on the west side, extend beyond the width of one archway of the blind loggia, incorporating a column; they are supported by two stone corbels, single on the south-west side and divided into two levels on the north-west side. The landings on the east side are not so wide and fit within one archway of the blind loggia; as such, they have just one corbel made up of two stone elements with quarter-circle profiles. It is not clear why wider landings were constructed on the west side,

Figure 2.20 Test highlighting the column of the south-east flying buttress incorporated into the brick masonry.

Figure 2.21 Dome, drum and level of the clerestory. The staircases on the extrados of the south flying buttress are visible.

Figure 2.22 Staircase at the top of the flying buttresses.

given that in all cases, the flying buttresses and the door providing access to the internal stairs of the crossing are positioned within the same archway of the blind loggia.

The corbels of the landings required the simple chiselling of the shaft of a column and one base, respectively, those on the south-east and north-west (Figure 2.23). Taking into account that the corbels were put in place before the columns in that they were inserted at the same time as the surrounding masonry, in all likelihood, there was a slight error in the positioning of the corbels which made these chisellings inevitable in that the position of the columns was fixed by the span of the arches of the blind loggia.

The initial section of the lower part of the staircase created on the flying buttresses is wedged into the sides of the transept. This part of the stairs was positioned in a masonry area close to the eaves and the corners between the transept and the longitudinal body (Figure 2.25). Creating the staircase inside the wall, which was not particularly thick,

Figure 2.23 (a) Landing of the staircase on the flying buttresses. (b) Detail of the corbels supporting the landing at clerestory level.

Figure 2.24 View of the crossing with the staircases linked to those of the flying buttresses.

involved altering the setting of the external pilasters. Here, in fact, the wall of the staircases has an overhang and juts into the spaces between the pilasters, varying their height.

The stairs on the flying buttresses are mentioned in a document dated 1597 in order to specify the location of roof renovation work, carried out by the master Pasquino di Mariano (Casini, 1983, 514 and following). Other documents from the second half of the 18th century concern renovations to the stairs of the flying buttresses (Casini, 1983, 691–694, 696).

Figure 2.25 (a) Detail of the door that connects the staircase on the flying buttresses to the one inside the crossing. (b) The initial section of the staircase on the flying buttresses was created in the outer walls of the transept.

2.3 The pendentives

The first elements of the drum that start the gradual transition from the rectangular base to the elliptical plan of the dome are the pendentives (Figure 2.14).

During the first analysis of the point cloud, a difference in the height of the springers was noted, marked by a stone cornice with painted dentils running around the entire perimeter of the crossing which is therefore inclined. The south pendentives are positioned approximately 15 cm lower than the northern ones.

The plan measurements of the pendentives show a clear difference, on average 70 cm between the two sides of each pendentive. This difference is due to the position of the under-lying structures: the intermediate pillars of the matronea dictated the size of the longer sides of the pendentives: the support of the frontal arch of the pendentives positioned on the longer sides of the crossing in fact corresponds to the lower pillars of the matroneum (Figure 2.16). In turn, these pillars coincide with the large columns at the level of the church, thereby form-ing the constructive device that transmits weight from the dome to the ground in the most simple and effective way (Figure 2.8). The shorter sides of the pendentives instead rest on the masonry that sits on the transverse arches of the domed space (Figure 2.26).

The scalene form of the pendentive plan, as we shall see later on, is the result of the choice to create an octagonal drum with elongated north and south sides just above the pendentives, while in each western and eastern part, there are three sides of similar width.

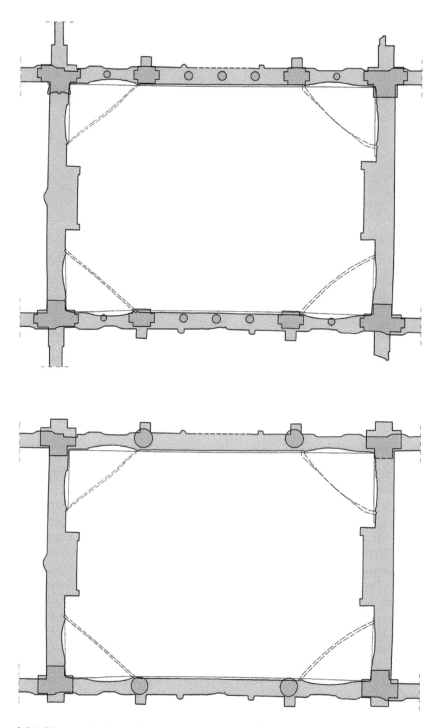

Figure 2.26 Plan at the level of the pendentives overlapped with the plan of the church (b) and with the plan of the matronea (a).

The pendentive is framed by a very large arch, with a span of roughly 5.50 m, which marks the start of the octagonal part of the drum. The actual pendentive, by means of its triangular plan, connects the diagonal frontal arch with the corner masonry of the crossing below (Figure 2.27).

Differences can be observed between the pendentives: the western ones are divided into two parts with different masonry structures, also corresponding to a different geometry, conical in the inner corner and "roughly spherical" in the remaining part, as Sanpaolesi observed when analyzing the north-west pendentive (Sanpaolesi, 1959, 207). The point cloud, however, highlighted the two geometries, not only in the western pendentives but also in the eastern ones where there are no differences in the wall structure.

In order to highlight the geometry of the pendentives, inside each of them, contour lines rising from the springer moulding at intervals of 20 cm have been indicated (Figure 2.28). The pattern of the contour lines effectively highlights the squinch in the innermost part,

Figure 2.27 Pendentives and summit of the staircase of the crossing.

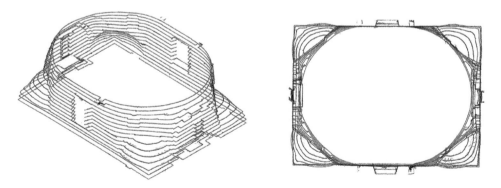

Figure 2.28 Plan and axonometric projections of the drum; the contour lines show how the form of the interior surface varies.

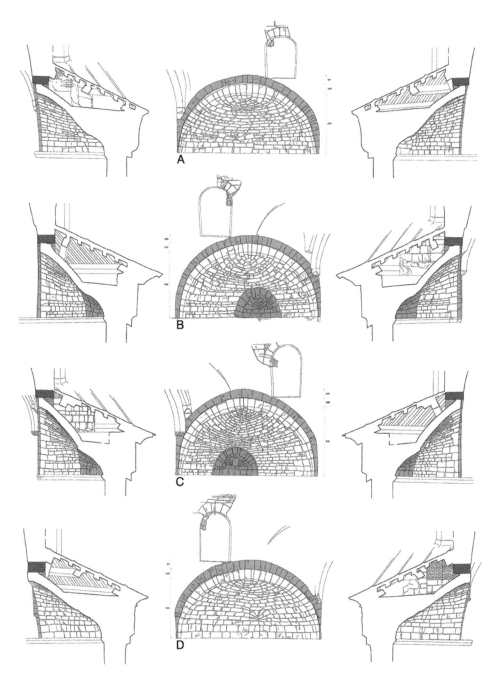

Figure 2.29 Elevations and sections of the south-east (a), south-west (b), north-west (c), and north-east (d) pendentives; the arches at the edge of the pendentives (light blue) and the interior squinches (brown) in the pendentives on the west side are highlighted.

which is distinct from the remaining part which is half-dome shaped. It can also be seen how the contour lines are closer together at the sides and more spread out in the central area, where the inclination of the pendentive reduces.

The elevations and sections of each pendentive show that in the western ones, there is a clear division between the two aforementioned geometries, corresponding to masonry sections separated by an archivolt of ashlars, which juts out in the north-west pendentive and is coplanar with the rest of the facing in the south-west one (Figure 2.29). In the eastern pendentives, on the other hand, this division shows no traces of interruption in the masonry structure, where the stones are arranged in continuous rows extending from the archivolt up to the innermost corner area. It is therefore plausible that as the pendentives were built, the builders experimented with how to construct them (Figure 2.30).

This difference between the four pendentives can be likened to a highly probable evolution in how they were constructed. The precision of the internal squinch raises the possibility that the eastern pendentives were constructed first, followed by the western ones which are more technically advanced.

The north-east pendentive has unique characteristics in the front, which are clearly evident in the longitudinal section: the frontal archivolt and the arch framing it rest on intersecting levels with different inclinations, while in the other three pendentives, these same elements have the same inclination. More specifically, in the north-east pendentive, the two adjoining arches coincide for an initial stretch; then the external arch rests on the projecting corbels and rises on an essentially vertical plane assuming a position at the top that is set back with respect to the archivolt of the pendentive. Indeed, in the north-east pendentive, the forward inclination of the archivolt is particularly pronounced; therefore, variations during construction prevented the external arch from tracing the same line. It can be observed that the greatest inclinations of the fronts of the pendentives occur on the east side: in the north-east and south-east ones, it is respectively around 40 and 25 cm versus 8 and 9 cm in the north-west and south-west ones. This is a further aspect, in addition to the geometries and wall structures, that differentiates the pendentives of the west and east sides of the crossing.

As regards the extrados of the pendentives, we have a description from the late 19th century:

> At the base of the dome are four lunettes in which there are oil paintings of the Four Evangelists, which day by day can be seen to peel away due to damp spots that appear

Figure 2.30 Pendentives on the west side.

on the wall. Having climbed into the low attic between the aforementioned lunettes and the roof, I was able to ascertain that water filtration does indeed occur and specifically in the lines where the roofs of the central nave and the walls of the dome meet, and this is also due to the far from perfect flashing that now exists, formed as usual by a curved sheet of lead fixed to the wall by a cement mortar. The futility of using cement in contact with lead and a smooth wall is immediately clear. With temperature changes, the metal shortens or lengthens detaching from the cement mortar, and a significant crack always occurs, allowing the garret water to penetrate: the flashing should be made of a sheet of lead or copper in a manner similar to that indicated above [...].

As the lunettes are constructed in tuff, they fully absorb all the water that trickles over them, with clear damage to the paintings executed on the surface facing the inside of the Church.

So in all circumstances, to prevent any filtration from damaging those paintings, I prefer to cover the rough part of the lunettes facing the attic with strong lime concrete in turn coated with a layer of Portland cement plaster; this would mean the water can no longer be absorbed by the material of which the lunette is constructed as it would readily run off the watertight and smooth surface.

(Casini, 1983, 955–956)

Today, the extrados of the pendentives is concealed by a mortar screed. Nevertheless, the extension of the actual shell described in the 19th-century text is considerably smaller than the visible surface of the intrados. The base of the pendentives is in fact an integral part of the rectangular perimeter wall at the base of the drum. Moreover, taking into account the thickness of the pendentive vault, roughly 25 cm, and the inclination of the vault itself starting from the springer, it is clear that the part of the intrados of the pendentives directly connected to the walls extends for around 1/3 of the vertical development of the pendentive; only the remaining part is the actual shell of the pendentive (Figure 2.31).

The visible masonry structure of the pendentives also allows us to study the particular placement of the ashlars. After the first roughly horizontal rows, from the ashlars of the frontal archivolt, the masonry structure continues on the inside with inclined rows, which are frequently double and gradually reduce in height. This latter aspect seems to be dictated by the geometry of the pendentives and in particular by the difference in the curvature between the frontal arches of the archivolt and the longitudinal profile of the dome-shaped area. This latter, highlighted by the section of the pendentives, has a less pronounced curvature than that of the archivolts; therefore, with respect to the same springer, of the two curvatures, the longitudinal one is shorter. This means that, if the aim is to create continuous rows, the dimensions of the ashlars of the archivolts must be gradually reduced in height when the rows reach the innermost part of the pendentives, thereby creating inflected rows. It should be added that the ashlars of the archivolts accompany the inclination of the rows, which is gradually more accentuated as they approach the top of the vaulted structure (Figure 2.29).

A comparison of the four pendentives reveals differences in the arrangement of the ashlars: there are rows in continuity with the frontal arch, as seen in the south-east and north-east; there is a mixed situation, with rows that divide with respect to the single wedge-shaped ashlar of the frontal arch, running from the aforementioned archivolt to the rest of the pendentive in two or three rows, and others that are in continuity with the ashlars in the arch, in the south-west pendentive; and lastly, the north-west pendentive stands out for the smaller and

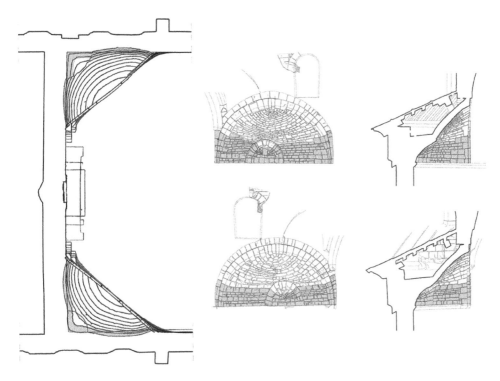

Figure 2.31 The plans, sections and elevations highlight (brown) the part of the wall of the
pendentives probably connected to the adjacent masonry of the drum; at the
level above, the pendentives have an autonomous shell structure.

more well-finished ashlars used. The distinct feature of this pendentive lies in the fact that
although the connection is always formed by the ashlar of the frontal arch, the subsequent
rows that connect with it divide into several rows, which are consequently smaller and make
it easier to manage and control the construction with inflected rows, which curve slightly
like a fan. This particular masonry structure essentially concerns the part of the pendentive
made up of the shell.

On the basis of all the information described, it is likely that the north-east and south-east
pendentives were the first to be built, and the west-side pendentives, with the south-west
and north-west pendentives, were tackled later, probably after a period in which the building
phase between the east and west sides was interrupted, which would explain the different
morphology in the construction of the corner of the pendentive.

2.4 Form and structure of the drum

The transition from the rectangular plan of the crossing to the irregular octagonal one, apparently achieved by means of pendentives, is actually made up of a complex system of arches
with connected springers. The visible masonry structure of the archivolts of the pendentives
and that of the arches that frame them show how these structures, before rising as independent elements, share from one to four ashlars at the springer (Figure 2.32). These are special

Figure 2.32 Surveys of the pendentives with the system of cracks (in red).

pieces that were not made of panchina livornese sandstone like the rest of the pendentive but limestone to make these construction junctures more resistant. The south-west pendentive is an exception as the common element is a monolithic part of the stone cornice at the springer of the two arches. The arches have Florentine archivolts of a similar height and appear to have discontinuous contact: the arch of the front of the pendentive seems to have been used as centring for the arch above; between the two structures, thin inserts of mortar with stone fragments can be seen (Figure 2.33).

The two arches, however, are of clearly different sizes, as shown in the sections. The archivolt of the pendentives is approximately 30 cm thick and comes into contact with a limited area, reduced to just 20 cm, of the intrados of the arches above (Figure 2.29). The arches above are much more massive, and their width varies from 63 to 69 cm. An inspection of the small garrets behind the pendentives revealed further important construction elements that form part of the original structure of the drum. The inspection and the survey of these structures were particularly complex due to the cramped garrets, created by the sloping roofs of the central aisle of the longitudinal part of the building. The garret area is in fact limited to the parts close to the corners of the east and west sides of the octagon of the drum (Figure 2.34).

Figure 2.33 Detail of the archivolt of the pendentive and of the arch above.

The same elements that appear in the crossing are also partly visible, that is the highest part of the extrados of the pendentives, here entirely covered by mortar, and a part of the archivolt and of the intrados of the arches above.

Above the arches that frame the front of the pendentives, there is also meticulous stonework coeval to an additional arch, flanked by the one below but with a wider arch. These side-by-side arches on the external side of the drum abut corner pillars corresponding to the corners of the octagonal part of the drum. The arches, masonry structures and pillars that can be seen in the garrets, all made of panchina livornese sandstone except for one pillar partly constructed of bricks and one pillar that could not be examined as it is covered with mortar, are of coeval execution. The close building phase connection between the wall and the external arch of the drum is particularly clear in the garret of the south-west pendentive, where these elements appear to be perfectly interlocked (Figure 2.34a).

The plan executed just above the springer of these arches further clarifies the relationship with the remaining constructive elements of the drum (Figure 2.35). Meanwhile, it should be said that on the sides of the pendentives, the external arches are cut diagonally by the roof slopes, which slope down towards the arms of the transept (Figure 2.36). So in the visible parts on the outside of the drum, the facings of the arches and the lunettes below are made almost entirely of two stones of marble, San Giuliano marble combined with green marble from Asciano (Sanpaolesi, 1975, 201); in addition to this, in the innermost part of the intrados of the arches, there are some ashlars made of panchina livornese sandstone. The plan shows that, in each of the diagonal walls, the marble parts of the arches are placed in continuity with the panchina livornese sandstone sections of the garrets. In other words, the same arch uses different materials in the visible parts and the parts incorporated into the garrets. The sum of the internal arches, those into which the pendentives fit, flanked by outer arches, determines at this level the overall thickness of the oblique sides of the drum, 107 cm on the eastern sides and 93 and 99 cm on the western sides.

The arches on the north and south sides, entirely visible and made of marble, stand on pillars also made of marble. The same material makes up the parts of the drum visible above the arches, up to the base of the loggia (Figures 2.37 and 2.38).

Figure 2.34 (a) Detail of the external arch of the drum connected to the adjacent masonry. (b) Garret of the south-west pendentive, the initial part of the external arch of the drum and the inner side of the arch framing the pendentive can be seen. (c) Garret of the north-west pendentive; from left, the arch above the pendentive, springer of the external arch of the drum, pillar.

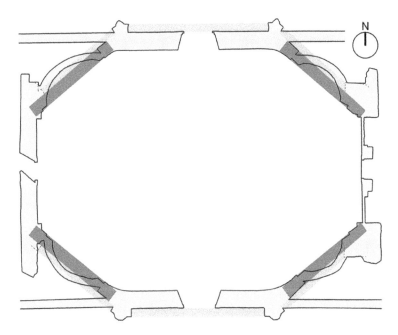

Figure 2.35 Intermediate plan of the pendentives, the arches that frame the pendentives (brown) and the external arches of the drum (cyan) are highlighted.

Figure 2.36 The dome and drum with the external arches partially covered by the roof of the nave.

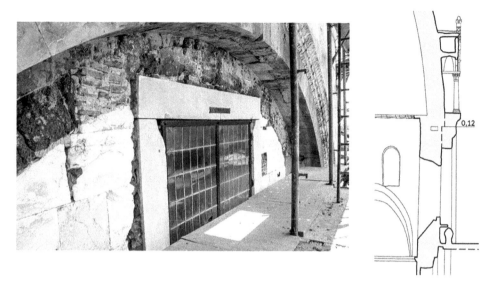

Figure 2.37 Part of the section of the drum and image of the southern arch, the dome
slightly overlaps the external arch.

Figure 2.38 Part of the section of the north side of the drum and photo of the external
arch, which is overlapped by the dome for a short stretch.

The wish to make the external facings of marble and those that are not visible, or for
the most part less visible, in panchina livornese sandstone, created distinct transition areas
between these materials. The transition between the two materials can be seen close to the

diagonal line of the roofs of the diagonal sides of the drum and in some cases at the intrados of the external arches. At the level of the summit of the matronea roofs, above the coffering, the first row of the marble facing that continues into the windowed side of the central nave above can be seen. In the marble apsidal conch, panchina livornese sandstone was used for the internal wall of the small crowning loggia, which is in fact hardly visible.

The fact that the drum was originally created with external arches, as a single architectural and constructive system, is further confirmed by the relationship between the drum and the dome above highlighted by the cross section: in fact, without the external arches, the extrados of the dome would have protruded by about 10 cm on the north and south sides, which would have been unacceptable in construction terms (Figures 2.37 and 2.38).

The tests carried out in the lunette of the outer north arch of the drum showed that the current rectangular opening created later was preceded by a trifora: on the inner and outer sides, the central arch and part of the side arches on just the external side have been highlighted. In this case, the external marble arch skirts some ashlars of the trifora, embedding itself into the wall and proving the coeval execution of these elements; on the inside, above the central arch of the trifora, there is no trace of an arch. Therefore, the external arch was interlocked with the adjacent wall without completely crossing it, just like a real discharging arch.

A completely different construction choice was made on the east and west sides of the drum (Figure 2.39). There is a discharging arch, namely an arch that extends for the entire wall thickness of the drum. Both arches, on the inside of the crossing, rest on slightly projecting corbels; the arch to the west is almost completely infilled, while the other one opposite has been closed up with a thin wall stiffened in the central part by massive pillars that, above the door interposed between them, form a single masonry unit (Figure 2.40).

The plan of the drum has another interesting dimensional aspect. The skew arches at the front of the pendentives and those on the east and west sides have roughly the same span, with size variations ranging from 5.20 to 5.45 m, while the arches on the north and south sides have a considerably larger span of 7.40 m. This created a very elongated octagon, also dictated by cogent constructive reasons, as the pillars of the arches rest on the top of the walls that flank the nave of the transept arms (Figure 2.26).

This irregular octagonal form is essentially maintained on the outside up to the summit of the drum, except for the adaptations on the west and south sides which we shall focus on below. Instead, several constructive solutions were employed inside the crossing to obtain an ellipse at the springer of the dome.

It should be stated that a comparison of the rectangular plan just below the pendentives and the elliptical one at the base of the dome shows that they have the same dimensions on the north-south axis. In the middle areas of the longer side of the crossing, the walls of the drum are essentially vertical, while in the remaining parts, that is the corner areas and the east and west sides, different construction strategies had to be put in place in order to extend the walls inside the crossing to obtain the ellipse at the base of the dome.

This change to the profile of the wall starts to occur in the pendentives on the east side. The wall of the front arch does not maintain a vertical plane as it rises but tends to protrude forward. So, with respect to the starting position of the springer, in the north-east and south-east pendentives, there is a significant shift forward, while in the westerly section, the arches have a slight inclination.

On the shorter sides of the crossing, at the level of the garrets of the central nave, the first forward shift is seen with the overhanging arches on corbels partially embedded into the archivolts of the arches that frame the pendentives. Just above, on each side, there is a further overhanging

Figure 2.39 East side of the drum; the following are highlighted: the pendentives (yellow)
with the arches above (brown), the central discharging arch (light blue) on corbels (green), the arch above (blue), and the asymmetrical arches (red).

arch. On both sides, even if more accentuated on the east side, the arches project forward as
they rise (Figure 2.41).

These forward inclinations of the arches are accompanied by a similar trend of the walls
between them and the pendentives and those that flank the pendentives themselves on the
north and south sides.

A test, which removed part of the plasterwork of the two arches on the east side of the
crossing and of the wall interposed between them and the north-east pendentive, revealed a
very precise masonry structure: all the construction elements are made of panchina livornese
sandstone, carefully squared and laid with thin joints (Figure 2.41).

The overhang of the wall to obtain the elliptical plan was therefore obtained through
arches, sometimes also inclined to accentuate the extent of the shift towards the inside of
the crossing, or by creating inclined walls. The greatest overhang was necessary on the
shortest sides of the crossing. So diagonal arches were added in this area, which on one
side rest on the central areas of the pendentives, and on the other on the upper arch of the
east and west walls. Each of the diagonal arches has springers at very different heights, of

Figure 2.40 Infilled arches on the east and west sides of the drum.

about 1 m; the highest one was in fact required as it had to be placed above the arches of the east and west walls. The lowest springer seems to be linked to the choice not to place the arch at an intermediate level of the diagonal walls of the octagon, which are of moderate thickness (approximately 68 cm). It must have been considered more effective to place the support just above the structural system formed by arches at the base of these walls and by the pendentives below. The asymmetric arches, in that they are skewed with respect to the walls they are inserted into, have an intrados that in part is roughly triangular if projected on a horizontal plane, but most likely it is only a small part of an arch incorporated into the masonry of the drum (Figures 2.42 and 2.43).

The walls above these asymmetrical arches and those that rise from the remaining parts of the drum were still built with an overhang in order to obtain an elliptical drum at the base of the dome. In order to support this trend of the internal surface of the drum, the jambs and archivolt of the arched windows positioned just above the pendentives are necessarily inclined. The arches of these openings and the adjacent masonry, on the basis of tests carried out, are made of panchina livornese sandstone like the remaining parts of the drum (Figure 2.44).

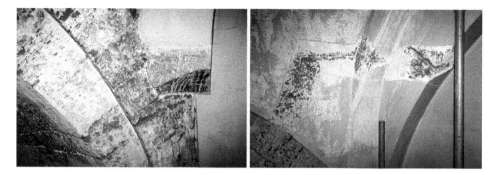

Figure 2.41 (a) Detail of the arches of the pendentives and of the corbel of the discharging arch of the west wall. (b) Test highlighting the homogeneous characteristics of the masonry facing of the drum arches.

Figure 2.42 The asymmetrical arch between the south-east pendentive and the relieving arches of the east wall of the drum.

The form of the internal surface of the drum, from the base rectangle to the ellipse at the top, is therefore based on a complex construction system formed of overhanging and skew arches, with asymmetrical supports, and inclined walls (Figure 2.45). This resulted in the creation of a sort of large wall corbel on the east and west sides, in the areas of the pendentives and, partially, on the north and south sides. This corbel gradually extends for the height of the drum up to around 8.30 m and reaches a maximum overhang of 95 cm on the side nearest to the apse (Figure 2.46).

This complex set of elements and construction solutions reflects extraordinary execution skills which, despite the frequent skewed connections, still involved the careful working of the stone material, laid with squared ashlars and thin joints (Figure 2.47).

The tests carried out just above the arched windows of the drum provided further information on the materials and construction elements. On the north and south walls of the crossing,

Figure 2.43 SBAAAS PI 5558, Springer of the asymmetrical arch on the pendentive.

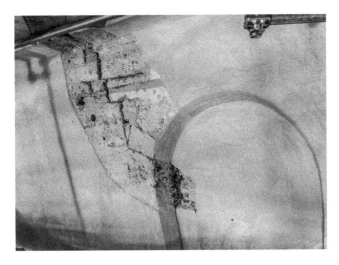

Figure 2.44 Detail of the window above the pendentives; the cracks spread from the opening to the top of the drum.

beam slots were identified which allude to a work plan most likely prepared for the construction of the dome, considering their position close to the top of the drum (Figure 2.48). Just above these slots is a cavity longitudinal to the north and south walls which must have accommodated a wooden tie. The tie skirts the arched windows of the drum and is positioned below the central area of the thickness of the dome. It should in fact be considered not so much hooping at the base of the dome but rather an element that helps to make the top of the drum less deformable as the masonry shifts from stone to brick. The tie marks the level

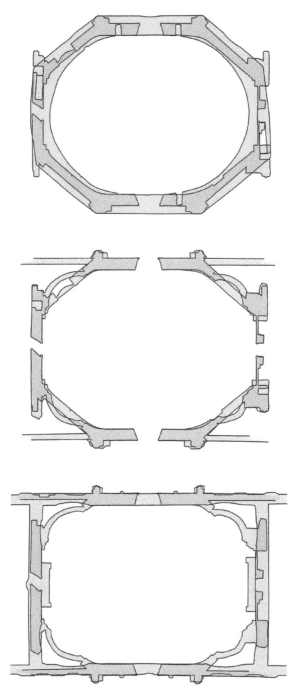

Figure 2.45 Overlapping between the plans of the drum at different levels; rectangular drum and intermediate level of the pendentives (c), intermediate level of the pendentives and octagonal part (b), and octagonal part and summit with elliptical profile on the inside and octagonal on the outside (a).

Figure 2.46 Inclination of the arches and of the walls on the east side of the drum.

where the brick wall starts, which then completes the drum by continuing the inclination of the masonry until reaching the internal crowning balcony made up of a series of highly segmental small brick arches which alternate with stone corbels.

With respect to a similar complex construction, the geometric control capacity demonstrated by the builders in the shift from the rectangular base to the elliptical top of the drum should also be pointed out. The elliptical profile of the drum, with longer and shorter axes, respectively, of 16.03 and 12.54 m, if compared with an ideal ellipse drawn with three centres, as proposed by Sanpaolesi (1959, 217), records average deviations of no more than about 10 cm (Aita et al., 2017, 44–45).

2.5 The stairs in the drum: the constructive connections between the dome and the loggia

The loggia at the base of the dome can be reached by two staircases created in the east and west sides of the drum: they start from the garret of the central nave and, via a single ramp, lead to beneath the vaults of the loggia. The two staircases have a similar position; they are accessed from the corners of the octagon of the drum, one between the west- and north-west sides and the other between the east- and south-east sides. The position is not accidental in that it meant the entrance to the stairs could be created in the masonry between the haunches of the arches that converge on these corners of the octagon. Moreover, the height of the entrance to the stairs resulted in the creation of ramps that, in the haunch area, skirt the extrados of the arches on the east and west sides of the drum (Figure 2.49).

These stairs cross the part of the drum that changes from an octagonal to an elliptical shape. In construction terms, this area is highly complex and should also be examined in relation to the position of the dome. The overlapping of the plan view of the springer of the dome and that of the drum shows that, in the central area of the longer sides of the crossing, the dome sits directly on the masonry made up of walls that rise from the matronea, integrated by the external arches

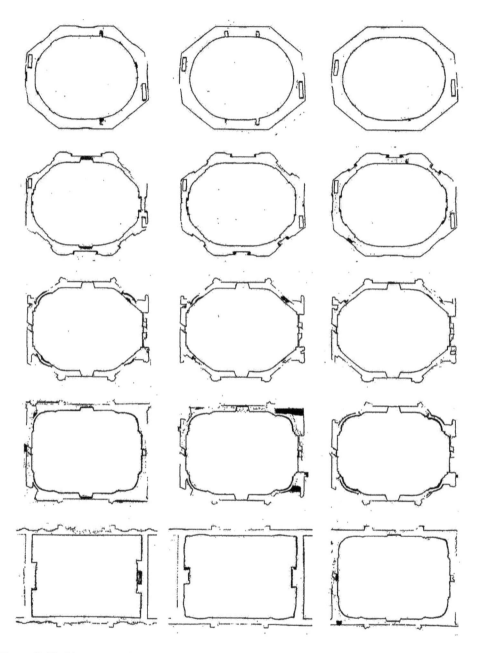

Figure 2.47 Changes to the interior profile of the drum, from the rectangular perimeter at the base of the pendentives to the elliptical form at the springer of the dome.

of the drum. On the shorter sides of the crossing, the dome is instead entirely positioned over the wall corbel part of the drum overhanging the crossing, except for a slight overlapping with the west wall of the octagonal level of the drum itself (Figure 2.50).

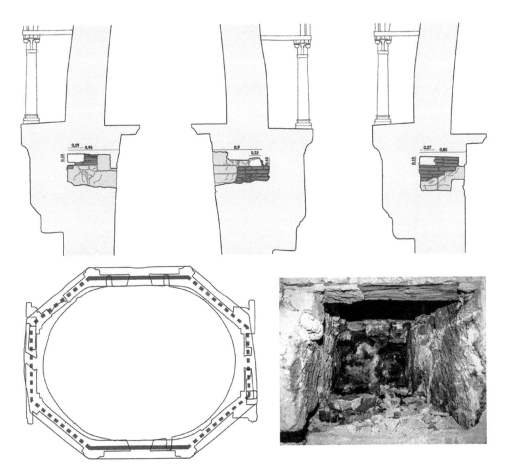

Figure 2.48 Detail of the areas where beams were inserted and used to create a work platform to construct the dome, indicating the position of a wooden tie (in yellow) positioned inside the masonry; plan showing the position of the wooden
tie on the north and south sides (brown), the position is hypothetical in the
dashed parts; photo of a support slot for a wooden beam.

It is worth further exploring the relationship between the stairs, each in its specific constructive context, and the dome. The stairs are of similar width, roughly 55 cm, and on the
side of the garrets, they are flanked by a wall of around 30 cm. Both were constructed necessarily taking two conditioning elements into account: the width of the arches below them and
the fact that the landing area is at the top at the edge of the extrados of the dome (Figure 2.51).

These conditions should have been met by the builders taking into account a difference in
context: the east staircase was built on an arch approximately 135 cm wide and the west one
on an arch 110 cm wide. This meant the east staircase was slightly repositioned to the outside
in the area skirting the dome, while on the opposite side, a more obvious repositioning was
required.

These adaptations to the staircases made the central part of the masonry thicker on the
external east and west sides, which are the only ones on the octagonal perimeter of that

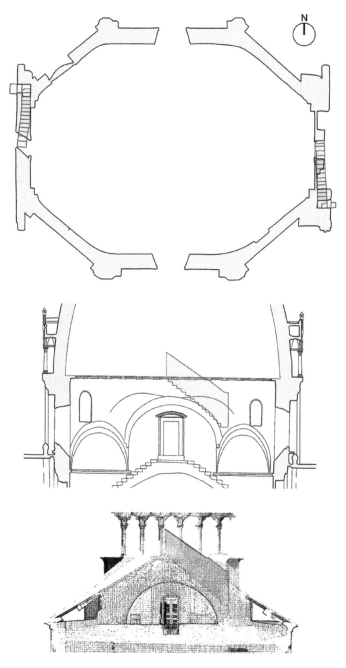

Figure 2.49 (a) Plan indicating the staircases of the drum (light blue and green). (b) Section with the position of the staircase indicated in transparency. (c) Section obtained from the 3D model of the laser scanner.

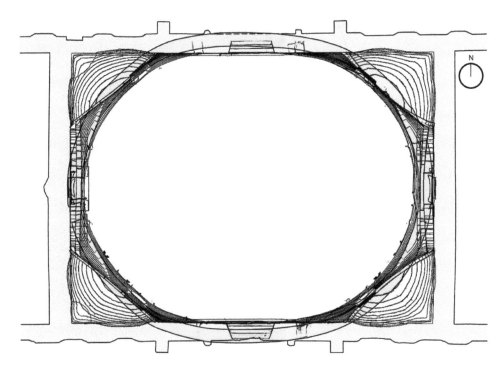

Figure 2.50 Overlapping of the drum and dome (in green); most of the dome rests on the overhanging structures.

curve, thereby creating a thickening. On the east side, where the repositioning is less, the brick masonry lies in overhanging rows, and at the top a section of the marble cornice at the base of the loggia above is wedged into them. On the west side, in order to obtain a greater increase in the wall thickness, a brick segmental arch on stone corbels was created, above which is a brick wall crowned by a section of marble cornice at the base of the loggia. Brick masonry continues up the sides of the arch on corbels, gradually connecting the overhang of the arch with the rest of the wall, creating an extended curved profile in the highest part of the outer side of the drum (Figure 2.52).

The construction of the part of the staircase corresponding to the highest steps implies that the position of the dome had been fixed at this point, and its construction had actually started. All this was carried out just above the level of the wooden tie, approximately 50 cm below the top of the drum. The masonry of the dome, which starts from the top steps of the staircase, has a vertical facing positioned in continuity with the cylindrical masonry at the extrados of the initial section of the dome, which rises up to almost the level of the loggia capitals. At this level, the external cylindrical form is interrupted with a wall offset, where the curve of the extrados starts. Sanpaolesi ascertained that this masonry is made up of whole bricks and had not been reworked (Sanpaolesi, 1959, 220, 223, fig. 40). In correspondence to the cylindrical section, the intrados of the dome has a continuous curve from the springer (Figure 2.52). The first stretch of the dome, up to the wall offset, is most likely made up of

Figure 2.51 Section of the east side of the drum showing the inclination of the staircase and view of the stairwell.

masonry arranged in horizontal rows, which gradually overhang on the inside in order to create the curvature of the intrados from the springer upwards.

A similar situation is seen in Florence Baptistery, where the first section of the octagonal dome is made up of masonry arranged in horizontal rows, with a vertical extrados and curved intrados (Giorgi & Matracchi, 2017, 194–195). The first stretch of the dome inside Siena

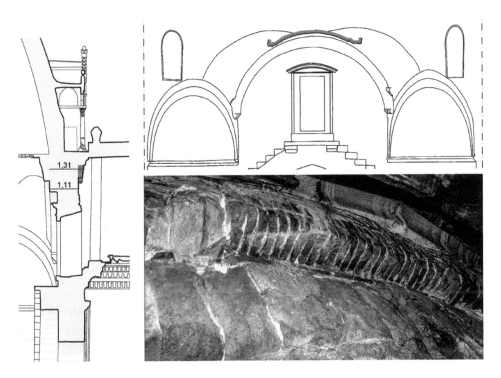

Figure 2.52 Small overhanging arch on the west side of the drum, highlighted in red in the section and elevation.

cathedral is also formed of overhanging horizontal rows (Giorgi & Matracchi, 2010, 44, 50), in this case with the extrados parallel to the intrados.

Other useful information can be deduced from the masonry visible in the stairwells of the drum of Pisa cathedral. In the east one, on the inner side of the initial section of the stairs, there are carefully laid large stone ashlars which, in the entrance landing area, rise up to the roof of the stairwell, made of transverse stone lintels. On the opposite side adjacent to the garret, there is a rather thin wall made of stone ashlars in the sections close to the steps, while bricks are used in the rest of the stairwell masonry. This material is used from the level of the entrance door to the staircase; it should be pointed out that here the stone facing of the inner side is interrupted by a wall offset, covered up by a thin adhering brick masonry. The characteristics of the stairwell walls on the west side are similar.

Considering the extension of the brick parts and the initial part entirely made of stone which goes as far as the roof, and taking into account the position of these stairs, it is clear that the drum, the stairs providing access to the loggia and the initial part of the extrados of the dome are elements strictly connected to and consistent with construction work that unfolded over time, perhaps the result of gradual decisions without real variations.

On the other hand, it seems highly unlikely that these stairs were created by breaking through the wall right at the edge of the sections where the dome rests on the corbelled part of the drum. With this operation, the corbel of the drum would have had almost no direct wall connections above the octagonal part of the drum, creating a deep breach, extending right up to the adjacent asymmetric diagonal arches and even to a part of the area below the dome. It would have been a variation made under extremely risky operating conditions, carried out moreover by builders who did not have detailed information on the internal structure of the drum, which was created by their predecessors. The stairs providing access to the loggia are therefore characterized as coeval to the drum.

The loggia at the base of the dome provides further elements of reflection on the construction work, considering in particular the construction relationship between the loggia and the dome.

The columns of the loggia are positioned on the irregular octagonal part of the drum, at varying distances from the extrados of the dome; the distances between the columns are just as variable. This last aspect in particular is important as regards the organization of the construction phases. The pediments of each bay of the loggia, formed by a trefoil with a triangular point and sloping sides adorned with crockets, are made of monolithic marble elements. Due to the variable width of the bays, the pediments are made up of special pieces (Figure 2.53). Pinnacles placed on the dosserets of the columns are interposed between the pediments.

The stability of the loggia is down to three different methods of anchoring the dome: a stone tie placed above the capitals, a further stone tie at the top of the pediments, and metal ties halfway up the pinnacles.

The stone anchoring at the level of the capitals is inserted just above a wall offset of around 10 cm, at the extrados of the dome, where the masonry of the dome starts to be approximately 60 cm thick. Sanpaolesi ascertained that this brick wall is two bricks thick and made up of bricks measuring $26 \times 13 \times 6$ cm (Sanpaolesi, 1959, 210). The stone ties resting on the capitals are, respectively, 10 and 30 cm thick and wide; they are inserted into the dome at average intervals of around 75 cm. The ties at the top of the pediments have a slightly smaller section, and, as they are in a higher position, they occur at shorter intervals due to the gradual

Figure 2.53 (a) Interior of the loggia of the dome and roof of the stairwell of the drum. (b) Detail of a gable of the dome loggia.

Figure 2.54 (a) Section of the loggia indicating the connections with the dome: the stone ties at the level of the capitals (blue) and of the top of the gables (green), the metal bars that anchor the dosserets (red). The stone ties at the top of the gables (b) and those at the level of the capitals (c).

reduction of the volume of the dome; just above them are metal ties anchoring the pinnacles (Figure 2.54).

The proximity of the insertion points of each of these three elements, the two stone ties and the metal tie, repeated for the entire perimeter at the base of the dome, created significant operational problems. To date, the tendency has been to think of the loggia as an addition to the dome; but in that case, the connections between the loggia and the dome surely would have been made by breaking through the wall, essentially creating on two levels – that of the ties above the capitals and that above the ties of the pediments almost coinciding with that of the metal ties – breaches that would have affected a significant part of the perimeter wall of the dome. Especially as such a connection made in a breach would have meant that the stone ties and metal ties had to be inserted much deeper into the brick wall two bricks thick in order to ensure the stability of the loggia with efficient interlocking. In other words, it would have required huge demolition work, even if carried out for temporally distinct parts.

In order to clarify how the connection between the loggia and dome was made, it would be useful to perform tests in the dome where the connection elements meet, taking into account that some of them may have been replaced during restoration work. Nevertheless, the problem of the relationship between the loggia and the dome will be examined in the next chapter, in a broader reflection on the construction phases and some key aspects of the construction of Pisa cathedral, with particular regard to the crossing area.

2.6 Piero Sanpaolesi and the restoration of the drum

In the 1950s, Piero Sanpaolesi, in his role as designer and superintendent, worked on the restoration of Pisa cathedral, in particular pursuing research on the dome and drum. The restorations considered here concern the pendentives and the part of the drum that was redesigned with painted wooden cladding around the third decade of the 17th century (Peroni, 1995, 17) (Figures 2.55 and 2.56).

Sanpaolesi, as his writings make clear, examined and accurately described the north-west pendentive, also making use of an advanced survey, capturing its salient aspects:

> [...] the pendentive is made up of a conical squinch that uses a precise geometrical formula to resolve the difficulty of connecting a surface that curves with a 90 degree angle, such as that of the rectangular at the base of the dome. On this conical squinch that ends with a frontal archivolt made of ashlars stands a second portion of the pendentive which is roughly spherical and softens the transition from the right angle of the cone to the arch that delimits it above, which in turn is a double Florentine archivolt, and marked along the whole front by a projection, by a line of shadow [...] the ashlars of the archivolt of the conical squinch below have been completely chiselled to give

Figure 2.55 SBAAAS PI 5027; View of the crossing before Sanpaolesi's interventions, with the balustrade of the stair-case still present.

Figure 2.56 SBAAAS Pl 5021, Lastra C 2938; (a) First tests in the north-west pendentives, the decoration of the staircase with balustrades and the claddings with painting panels of the arch below are still intact. (b) Detachment of the fresco by Cinganelli from the pendentive.

the impression, to those looking up from below, that the pendentive is not made up of two connected and collaborating types but rather that it is a real spherical pendentive.

(Peroni, 1995, 211) (Figures 2.56 and 2.57)

Returning to the matter fifteen years later, he added:

The Florentine arches [of the front of the pendentive] [...] create a bridge across the right angle, but not symmetrically with respect to this angle. The planimetric survey shows their actual position and also reveals the obliqueness towards the inside of the domed space, which somehow suggests that the drum bends as it shifts from the elongated octagon to the three-centred ellipse, such is the geometric layout of the base of the vault [dome] itself.

(Sanpaolesi, 1975, 227–228)

Sanpaolesi understood all the key geometric aspects of the pendentive, and as regards the innermost part shaped like a squinch, he provided information on the archivolt delimiting it, which he described as chiselled and with a surface continuous with the rest of the pendentive. The squinch of the north-west pendentive is now the only one that displays an archivolt with an extrados that projects slightly with respect to the adjacent wall; moreover, the ashlars of

Figure 2.57 SBAAAS PI 5560; Start of the removal of the plasterwork of the drum painted in bands of two colours.

the archivolt show no sign of the chiselling carried out to make the plaster stick better, while it can be seen clearly in the remaining stone facing.

This apparent incongruence of the archivolt can be seen in some photos of the restoration work found in the Photo Library of the "Soprintendenza di Pisa". The situation described by Sanpaolesi can be seen in the photo of the pendentive after the fresco by Cinganelli had been detached. Two subsequent images taken in 1958 document the subsequent phase of the deep chiselling of the archivolt ashlars and the relocation of thick slabs, each with the same profile of the chiselled ashlar below, which create the current projecting archivolt, with the fresh mortar between one ashlar and another still highly recognizable due to the fact that they had recently been laid (Figure 2.58). Sanpaolesi clearly thought that this must be the original morphology of the archivolt, which was altered in his opinion following subsequent renovations.

Some documents found in the archives of the "Opera della Primaziale Pisana" and of the "Soprintendenza per l'Archeologia, le Belle Arti e il Paesaggio" for the provinces of Pisa and Livorno allow us to clarify some further aspects of the restoration of the pendentives, which started with the north-west one.

In a report dated 29 January 1954, Sanpaolesi described some works that had become necessary due to the infiltration of rainwater and concerning the construction of wooden scaffolding suspended "at a height of over 30 m from the church floor" and the "detachment of a fresco enclosed in a niche making up the [north-west] pendentive at the Dome springer. The work includes fixing the colour of the frescoed surface and affixing light canvases to detach the work. It must then be brought to the ground and a special wooden frame constructed, following the profile of the pendentive, with the positioning and restoration of the fresco" (ASABAP Pisa, cartella F120 PD, fascicolo G28/3, 1955–1959). In a subsequent

Figure 2.58 SBAAAS PI **8310, 8313**; (a and b) Chiselling and reconstruction works in the internal part of the north-west pendentive (**1958**). (c) The internal part of the north-west pendentive reconstructed.

letter to Giuseppe Ramalli, president of the Opera della Primaziale, he stated that "This work, which mainly involves the detachment, restoration and relocation of one of the frescos, would require around four months of work which must be started no later than the month of April". In fact, only the works to construct the suspended scaffolding were completed on 8 June 1954 (ASABAP Pisa, cartella F120 PD, fascicolo G28/3, 1955–1959).

Further information on the chronology of these work phases can be found in the report drafted on 17 August 1956 by Bruno Farnesi, one of the skilled workers of the "Opera Primaziale":

We are pleased to inform you that the suspended platform assembled at the request of the "Soprintendenza" is still in place, under the Cathedral dome in correspondence to a "pendentive" frescoed with an Evangelist positioned in the north-west area.

This very ambitious platform, suspended and weighing over 800 kg and with a bearing capacity of 1000 kg is secured with hemp ropes and tie rods; it was to be used for a period of 4 months, but to date it has been assembled for a good 25 months.

The works carried out, as you are aware, were: the detachment of the fresco, which was laid on the matronea, where it still remains (this work was completed on 14.6.1954), the scraping of plastered areas, to make the tuff structures of the "niche" and walls visible, photographs and surveys (work that lasted until 18 December 1954) after which no work or interventions were recorded.

(Archive of the "Opera Primaziale Pisana",
in cartella 787, fascicolo 2826)

However, all these works were just the prelude to a further intense phase of restoration which also involved the lead roof of the dome (ASABAP Pisa, cartella F120 PD, fascicolo G28/3, 1955–1959) (Figure 2.59). This was the period in which the frescoes by Cinganelli

Figure 2.59 SBAAAS PI 7945; Brick extrados of the dome visible during the replacement of the lead roof.

in the other three pendentives were detached, according to a very widespread practice at the time, which seemed to be the most critical part due to the infiltration of rainwater and the presence of cracks.

It is clear that the project was defined after the suspended platforms made direct observation and the survey of the pendentives possible. So after having decided to investigate them all by detaching the frescoes, the condition of the structures was likely not deemed to be particularly serious as they decided against carrying out wall repairs in the pendentives to close the cracks. The masonry of the pendentives in fact appears to be intact, with the cracks still visible today, except for the intervention in the north-west pendentive, as noted, which has an archivolt overhanging the inner squinch.

The restoration work was not limited to the detachment of the Four Evangelists frescoes. This was only the start of a complex restoration, documented by Sanpaolesi in an article published in 1959 in the "Bollettino d'Arte", which concerned the whole drum. The area between the springer moulding of the dome and the staircases on the shorter sides of the crossing was restyled in the first half of the 17th century. Painted wooden boards anchored to the wall with metal cramp irons were used to clad the interior surfaces of the drum "to make it seem like an elliptical cylinder" interrupted only by the pendentives (Sanpaolesi, 1959, 211). It must have been how the 17th-century renovation was executed – boards anchored by metal brackets that did not require specific alterations to the underlying structure – that encouraged their removal, defined by Sanpaolesi as a "ripristino di liberazione" [clearance restoration] (Sanpaolesi, 1959, 213).

2.7 Architectural and conservation issues in light of the new acquisitions

In his fundamental monograph on Pisa cathedral, Piero Sanpaolesi raised a methodological problem in the study of architecture, mentioned only briefly here, and for the first time addressed significant construction aspects pertaining to the drum and dome, relating them to the construction phases of the dome (Sanpaolesi, 1975).

Sanpaolesi complained that too many studies still tend to reduce architecture to abstract patterns that lie outside the study of real buildings, or they approach them in entirely partial terms in order to confirm attributions to builders and construction chronologies. We must also highlight the interpretative need to relate the concrete constructive elements of architecture, considered as a source of central importance in philological work, to the archive documents (Sanpaolesi, 1975, 124–135).

Nevertheless, when Sanpaolesi claimed that cathedrals, as well as buildings of a more modest size, were necessarily constructed following an original design whose details were fully developed, he also made a significant simplification. This conviction considerably influenced some of his interpretations of the construction history of the cathedral, such as the dating of the dome which in his opinion should be considered part of the original design devised by Buscheto and already completed by the end of the 11th century. He recognized that the dome acts as a "generator of space and of the structures that support it, [where each part] is connected to the others due to the distribution and structural requirements". Sanpaolesi also identified a late 14th-century renovation during which the loggia at the base of the dome and the arches would have been built, eight in his opinion but in actual fact six, on the external sides of the drum (Sanpaolesi, 1975, 138–139, 181, 193, 208–209).

The topics addressed by Sanpaolesi can now be taken up and explored in-depth with new data using the very approach that in unequivocal words he suggested as follows: base the analysis and interpretation of architecture on a reliable survey, to be considered as the "first necessity for a well-founded study" (Sanpaolesi, 1975, 128).

The survey tools available today are much more sophisticated than those available to Sanpaolesi in the 1970s (Facoltà di Architettura, 1970). So the laser scanner survey performed especially for the new studies of Pisa cathedral allowed us to obtain the multitude of plans and sections that have gradually become necessary in exploring the many architectural and constructive aspects in-depth. Moreover, as already highlighted in the previous chapters, particular attention was also paid to the specific operating conditions in place during the construction phases in the creation and positioning of the complex structures that rise from the base of the crossing up to the springer of the dome.

The layout of the rectangular plan at the base of the crossing, delimited by cruciform pillars, is extremely accurate in terms of the geometric regularity and the size of the structures that delimit it. This contrasts with the irregular appearance of the arms of the transept and, in particular, the south one, where the west aisle gradually narrows up to its point of attachment to the basilica, where it assumes the width of the crossing (Figure 2.6). The marked narrowing of this aisle is due to the fact that it is delimited by an external skew wall. A similar unique feature makes it unlikely that the transept and the crossing stem from a single layout of the base structures; the definition of this part would seem to be the result of different construction phases.

On the shorter sides of the crossing, two large arches rise up and almost touch the drum. On the longer sides, above the trifora of the matronea, there is an apparently continuous wall, which actually has a discharging arch. This constructive solution, which envisaged the construction of an infilled arch from the outset, is highly important in that was also proposed again in the drum on the east and west sides in the skewed walls with four asymmetrical discharging arches (Figures 2.11 and 2.39).

The thrusts of the two transverse arches of the central nave are offset by flying buttresses with two archways, connected to the sides of the transept, which act as buttress walls. On the flying buttresses, just above the roofs of the side aisles, there are staircases that continue inside the crossing, where there are ramps supported by arches adjoining to the shorter sides; these staircases lead on each side to a common landing positioned at the height of the trusses of the central nave.

It should be pointed out how the sophisticated system of staircases linked to the flying buttresses, which most likely served to provide access to the roof for maintenance, create an interesting route that starts with the matronea, continues with an outdoor stretch that affords a close-up view of the clerestory and leads up to the entrance to the garrets of the central nave; from here, there is a close-up view of the drum and the dome and a precipitous glimpse of the crossing below (Figures 2.60 and 2.61).

This system of flying buttresses should not necessarily be related to the thrusts caused by a dome, in that it cannot be ruled out that a similar buttressing structure might have also been necessary with the construction of a groin vault crowning the space of the central cross.

On the other hand, the creation of the drum marked the start of a phase or architectural and construction experimentation initially concerning the pendentives, from which it can be inferred that the decision to build the dome could only have been made from this phase onwards.

Figure 2.60 The section highlights the continuity between the staircases on the flying buttresses and the interior staircases of the crossing.

Figure 2.61 View of the clerestory and of the baptistery from the staircase on the southwest flying buttress.

As highlighted by the study of the masonry structures, the pendentives on the east side have essentially homogeneous masonry, unlike those on the west side where the section close to the inner corner is delimited by an archivolt (Figure 2.29). A further difference is found in the frontal arches of the pendentives, which are markedly inclined towards the inside of the crossing on the east side and almost vertical on the west side.

The differences in the masonry structure of the pendentives, where the solution used on the west side should be considered the most advanced, derive from the geometric characteristics shared by all the pendentives. The longitudinal sections highlight that they are characterized by two curvatures: an external one that acts as a buttress and is moderately curved, and an internal one that forms a much more accentuated arch. The different profile in section corresponds to a further discontinuity on plan: the outer part of the pendentive is roughly dome shaped, whereas the inner part is a squinch with a triangular base.

This geometric differentiation in the pendentives is an important aspect to focus on. The rectangular wall perimeter just below the pendentives would have implied the construction of squinch pendentives, similar for example to those found in Siena cathedral, which however are not as deep. It seems that the thrust of the asymmetric skew arch which rests just above the pendentive was taken into account in the pendentives of Pisa cathedral. This created a pendentive which, like a sort of "strut", stiffens the part where the asymmetrical arch thrusts; it is no coincidence that the shell, right in the central part where there is an asymmetric arch, has a less accentuated curve that supports the shoring function (Figure 2.31). This "strut" rests on the inner squinch part of the pendentive, which is a sort of masonry corbel which slots into the corner, and as such, it is closely connected to the adjacent walls.

Experimentation during the construction phases involved the entire construction of the drum, in view of the complexity of the construction solutions implemented. It is made up of discharging arches, overhanging arches and arches that sometimes have an inclined archivolt. The discharging arches, namely arches constructed to be infilled at the same time, were used a great deal in the crossing space. They are found in each of the north and south walls at the level of the garrets of the central aisles of the transept arms; they occur on all sides of the drum at the level of the pendentives, where they are added just above the four asymmetrical arches which, as they are skewed, have an incomplete infill.

Discharging arches are also found in other extremely important buildings, such as above the triforas of the drum of Siena cathedral; according to the surveys conducted by Luigi Vanvitelli, they may have also been used in the cylindrical drum of St. Peter's basilica in the Vatican (Matracchi, 2013, 188–189).

Another distinctive construction solution of the drum of Pisa cathedral is the inclined walls, which act as a connection between the different constructive elements of the drum and made it possible to shape the form of the drum until the elliptical plan at the base of the dome was obtained. This harmonization of construction solutions led to the gradual definition of the drum, which started with a rectangular plan, transformed above the pendentives into an elongated octagon, which was in turn gradually shaped until obtaining a three-centred ellipse, with slight approximations. All this, above the six shorter sides of the octagonal part of the drum, created wall corbels with a considerable overhang where the dome rests. The wall corbel parts of the drum almost seem like cupular segments accompanied by the slightly set back dome above (Figure 2.46).

The outside of the drum has three arches in both the north and south parts. Without these arches, linked to the same construction phase as the pendentives, the north and south sides of the drum would not have been thick enough to support the entire dome (Figures 2.37 and 2.38).

The external arches overlap the top of the arched windows which skim each of the pendentives (Figure 2.62). It was thought that this unique feature would corroborate the nature of these arches as an addition. It should be said that the marble cladding of the lunettes of the

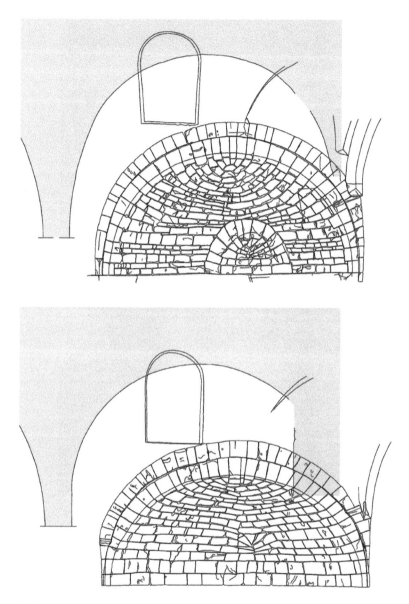

Figure 2.62 The archivolt windows of the drum in relation to the external arches highlighted in light blue.

arches is consistent with the jambs of the windows, also marble, which are now fully infilled.
This confirms that these openings coexisted with the external arches (Figure 2.63).

It should be taken into account that from the crossing, the arched windows could only
be glimpsed, thereby making it difficult to see their partial infill. The position of these windows was dictated by the constraints put in place by the pendentives and the asymmetrical
arches inside the drum. Furthermore, the arched windows were constructed with different

Figure 2.63 Detail of the jamb of the infilled window.

heights varying from 1.58 to 1.95 m, but even this significant dimensional gap is in fact not perceived.

After the pendentives, at the start of the octagonal part of the drum, two staircases were constructed and positioned on the east and west sides in order to create a connection between the levels of the trusses of the central nave and the loggia at the base of the dome. The staircases were positioned above the discharging arches, which are not wide enough to avoid the top of these staircases colliding with the position that the base of the dome would have assumed (Figures 2.49 and 2.51). So both staircases, albeit to a different extent, were raised with a slight deviation on the outside, obtained with overhanging brickwork. The need for this adaptation confirms that the decision to create the two staircases was taken after the pendentives were created, when the discharging arches on the east and west sides had been built, or at least their sizes could no longer be changed.

The drum was a particularly complex and consistent constructive device, but it was also the result of decisions made during the works. The construction site as a place for experimenting with alternative solutions and defining architectural and constructive structures left clear traces during the construction of the pendentives and the staircases of the drum providing access to the loggia.

The close constructive relationship between the drum, staircases and dome also involves the loggia itself, which, besides its connection with the dome achieved using dozens of stone ties and metal ties, appears to be coeval with the dome itself (Figure 2.54).

The archive documentation known to date concerns work on the dome between 1380 and 1381, and it should be considered that payments (1381) to the master mason Puccio di Landuccio to close the dome cannot be referred to the construction of the external arches of the drum as hypothesized (Tanfani Centofanti, 1897, 439–440). Other testimonies concern the development in 1383 of the model of the tabernacles for the loggia and its execution in 1383–1388 (Tanfani Centofanti, 1897, 343, 441–442). All these documentary references constitute, rather than the restructuring of the base of the dome with the addition of the loggia and arches that encircle the drum, the final outcome of long construction works marked by slowdowns and interruptions to the works, which is inevitable in construction sites of this magnitude.

Such a complex matter cannot be based on a presumed design traceable to the work of Buscheto, but rather it appears to be the collective result of the contribution of several builders and architectural and constructive choices made during the construction phases in different periods.

Sanpaolesi's conviction that the entire domed space was part of Buscheto's design and recognition of the dome's extraordinary Romanesque's character also became the premise that profoundly guided his choices in the restoration of the drum:

> All this [the inside of the drum] had disappeared under polychrome wooden decoration which concluded with a coloured and modilloned wooden cornice with scrolls, under which a dedicatory inscription painted in block letters on the plasterwork painted to imitate fake black and white marble completed the masking. This wooden cornice encircled the internal Romanesque cornice, and the whole of the seventeenth-century apparatus took away much of the value from the architecture of the dome, which could also have been made of wood, as well as added later. In any case, it was impossible to view and appreciate its forms and structure, and all critical analyses seemed to end when faced with this screen of painted boards, so the unique position of this building in the family of Tuscan Romanesque architecture was not perceived. As regards the structure of the dome, the most original and most valuable known in Tuscan Romanesque, it really was a commonplace setting made of painted wood.
>
> (Sanpaolesi, 1959, 207)

Sanpaolesi went on to say:

> There is no architectural validity to the wooden claddings which so decisively and with the petty ambition of mere apparent pomp altered the most put together dome of the entire Italian Romanesque world [...] So I proposed the restoration of the dome and the Opera del Duomo, with quick understanding, albeit with extensive and appropriate discussion, accepted the proposal.
>
> (Sanpaolesi, 1959, 213)

So the entire wooden apparatus that screened the inside of the drum was removed, leaving, at the end of the restoration, the situation documented by the surveys presented in this study. Today, the characteristics of the drum have changed again in that the frescoes of the Four Evangelists restored in 2018 were replaced in the respective pendentives, in a context which is now hybrid in that there are no 17th-century painted wooden claddings (Figure 2.64).

The fact-finding investigations on the complex structural device of the drum now provide a better understanding of the cracks that have spread over time, in particular in the area corresponding to the pendentives. These cracks, in some cases repaired several times as ascertained by Sanpaolesi (1959, 204), which above all concern the central part of the pendentives, then continue to the arched windows above, crossing the arches with more accentuated cracks, and then extend up to the dome.

In fact, it turns out that the asymmetrical skew arches generate an outward thrust from the vertical plane of the walls above the pendentives and, in the same position, the opening in the adjacent arched window helps to accentuate the loads. It should be pointed out that the diagonal walls of the octagon are not as thick as the remaining ones of the drum. It would seem that the builders, knowing all this, chose to create the pendentives with a half-dome shape with the intention of more effectively distributing the thrust to the masonry into which the pendentives are wedged. At the same time, they gave the central area of the shell an almost straight profile in order to create a sort of strut between the area that receives the direct thrust of the skew arch and the inner squinch part of the pendentive, which has a different curvature and acts as a masonry corbel connected directly to the adjacent walls.

The pendentive area was made more vulnerable by the presence of the arched windows, which were not surprisingly later infilled. Moreover, in this part, the complex system of

Figure 2.64 OPP-BCPSA source; Intrados of the dome with the frescos by Michelangelo Cinganelli relocated to the original position after the recent restoration works.

discharging arches characteristic of the drum was inevitably interrupted in favour of an arch that opens onto the pendentives.

The presence of the stairwells close to the two asymmetrical arches should be pointed out: in this case, in fact, these arches cannot count on a complete masonry connection with the adjacent wall of the drum. Moreover, the asymmetrical arches are found in the corbelled parts of the drum with the biggest overhang, where the full thickness of the dome rests. Considering this distinct construction condition, the asymmetrical arches generate a greater thrust in the south-east and north-west pendentives, where there is a slightly more wide-spread system of cracks (Figure 2.32).

The cracks in the pendentives may have formed early on and may have been one of the problems that had to be tackled by the committee of eight masters appointed in 1391 to examine the cathedral from the foundations to the top of the dome (Tanfani Centofanti, 1897, 344–345).

All these assessments provided a better understanding of the structural behaviour of the drum and enable us to identify the specific link between the potential condition of local vulnerability in the area of the pendentives and the consequent formation of the system of cracks.

The identification of potential local vulnerabilities should be considered key in under-standing structural efficiency in architecture, above all when dealing with complex buildings such as cathedrals, which are hardly ever threatened by systemic instabilities and almost always over the course of their history develop some localized conditions of greater struc-tural vulnerability due to genetic reasons linked to the construction, transformations, altera-tions brought about by restoration or weakening caused by earthquakes.

Chapter 3

Siena cathedral

3.1 The construction phases of the intermediate cathedral

Siena cathedral has an unparalleled, if not unique, construction history due to the ongoing architectural and constructive challenges that the builders had to face. Several large-scale transformations took place between the 12th and 14th centuries, overcoming configurations in the cathedral that were completed or for the most part already built in terms of their overall structure (Figure 3.1).

This was something entirely different from the usual, variable complexity of the genesis of large buildings, which were not based on a construction project as we would understand it today but are rather the result of the reflections and contributions of many builders who found a place of synthesis in the construction site. Here, the ideation, choice and production of materials and construction techniques came together during the building to create the architectural work.

But the incredible architectural synthesis of Siena cathedral does not stop here; a sense of the challenge that left behind signs of the risks, experimentation and failures in its stones can also be perceived. All of this makes it especially necessary to retrace some key passages of the cathedral's construction history in order to identify its countless unique features and more fully understand the reasons for the additional significant changes made to the layout of the structures after the long and incessant work carried out in the Middle Ages.

3.1.1 The remains of the original cathedral and the orographic characteristics of the site of the new church

The construction of the present-day cathedral transcended the previous one thought to have been built in the 11th century (Giorgi & Moscadelli, 2005, 49). Recent excavations of the latter revealed part of the semicircular apse built upon a foundation offset (Causarano et al., 2003, 156–158). There is little information on the origins of the new church, but it can be traced to the period 1120–1160 (Giorgi & Moscadelli, 2005, 62–63) (Figure 3.2).

The part of the apse found beneath the domed space of the present-day cathedral would seem to indicate that the building was not particularly large (Figure 3.3). Based on the apse, its position was not random given the underlying geological profiles: beneath a surface layer of inconsistent filling material lies a bank of conglomerate with a fault line dividing it into two parts, the western side lying roughly 5 m higher than the other (Pietramellara, 1980, 7–8). The fault line falls at the transverse diameter of the dome of the present-day cathedral (Rossi et al., 2003, 216–221).

Figure 3.1 View of Siena cathedral.

It is easy to understand why the original cathedral – the apse of which was found – was much shorter as it was built almost completely above the upper bank of conglomerate, while the present-day crypt and the Baptistery were set below this level, with their foundations on the lower bank up to the east front, which adjoins the present-day staircase to the east. According to the geognostic corings, the depth of the upper bank of conglomerate (taking the lower level of the crypt at 0.00 m as a reference) is between −0.90 and −4.20 m and the lower bank between −4.40 and −7.65 m; the overlying backfill could be approximately 4.00 m on the upper bank and as much on the lower bank, which probably sloped down to Vallepiatta (Rossi et al., 2003, 219–221).

It can also be observed that the level of the foundation offset in the apse of the earlier cathedral is roughly equivalent to that of the current cathedral square, in particular the area of the entrance to the Museo dell'Opera to the south of the transept of the present-day cathedral, which for the most part seems to have conserved the heights that existed in the 11th century (Figure 3.4).

The new cathedral, placed approximately in axis with the apse of the ancient cathedral and with its same orientation, started from the apse and the transept. This position on the one hand denotes the importance this site continued to have and on the other marks the coexistence of the old and new cathedrals to ensure continuity of worship. Usually, only one part of the older building was officiated until the one under construction could be used, at least in part. The remains of two earlier Franciscan churches were found underneath the Basilica of Santa Croce in Florence, the second one in particular is almost completely incorporated into the present-day church (Giorgi & Matracchi, 2011). Even more emblematic, because it is

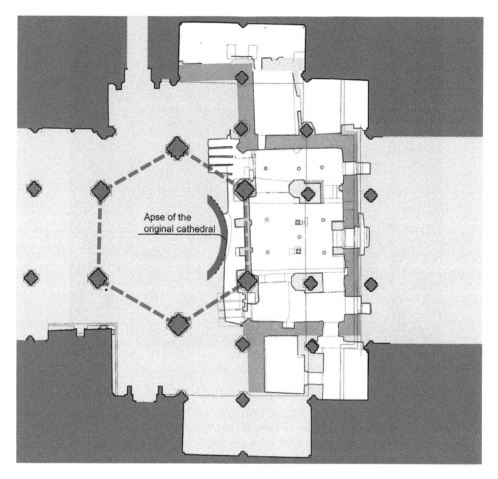

Figure 3.2 Plan of the crossing of the dome superimposed on the crypt and adjacent rooms; a part of the apse of the original cathedral is visible (red) positioned in line with the crypt wall (green) where scenes of the Passion are painted; the remains of the apse and transept of the intermediate cathedral (yellow) survive in the rooms of the floor below.

better documented, is the basilican body of Santa Maria del Fiore in Florence, which we shall return to later in the section of this book on the Florentine cathedral, where the earlier Santa Reparata church coexisted for a long time with the construction site of the basilican body of the new cathedral church (Giorgi & Matracchi, 2006a). The earlier cathedral of Perugia, which stood on a terrace from the Etruscan-Roman era, which over time has always been the most important part of the city (Matracchi et al., 2019), coexisted for a certain amount of time alongside the construction of the new cathedral which started in the first half of the 15th century. The earlier cathedral was positioned in the transept area of the new cathedral, so the construction works started with the aisles in order to delay the demolition and ensure continuity of worship (Matracchi, 2006).

Figure 3.3 Hollow space between the crypt and the remains of the apse of the original cathedral; the foundation offset can be seen at the bottom right.

3.1.2 The apse and the transept

Returning to Siena cathedral, the new apse stood on the lower east fault line and most likely on a slope, which posed specific construction problems. When digging the foundations, or to obtain rooms, the builders had to deal with the percolation of surface water which would run down steep slopes towards the excavations; this required the subsequent shoring of the ground which was gradually excavated. In this regard, the height differences between the foundation offset of the original cathedral and the present-day crypt are approximate, as is the much greater difference in height between the latter space and the baptistery, which would be built later. This must be taken into account when encountering transverse walls, for instance in the space uncovered by archaeological excavations, which also served to contain the mass of earth: examples could be the wall painted with scenes of the Passion reinforced by pillars and the parallel wall forming the interior wall of the present-day baptistery.

Figure 3.4 Longitudinal section of the crossing and adjacent bays; in the rooms below, the
level of the foundation offset of the apse of the original cathedral (red arrow)
and the adjacent part of the crypt (green) can be seen; the fault line roughly
aligned with the crossing is indicated in the subsoil.

The wing of the transept that extends towards present-day Via dei Fusari was also on an extremely steep slope; so on this side the back wall of the transept rises from a height that is around 7.30 m lower than on the opposite side (Haas & von Winterfeld, 1999b, tav. 18).

The Sienese builders must have developed a series of effective techniques to tackle such massive height differences, which the Cistercian experts called in to direct the cathedral construction work almost certainly did not know given that they came from San Galgano where they had not had to deal with excavations problems (Giorgi & Moscadelli, 2005); their expertise clearly concerned the structures above ground which, as important as they were, to modern eyes appear less demanding than the challenges faced in the containment of earth and bailing of the excavations. In any case, the decision to position the apse of the new cathedral close to that of the old cathedral, so as to limit its placement on the downward slope to Vallepiatta as far as possible, is indicative: the new apse is slightly setback with respect to the apse of the earlier cathedral. On the other hand, this particular orographic condition favoured the creation of a room below the apse.

The unique construction problems posed by a similar location are well highlighted by the pit foundations connected by arches, which as far as we can tell today were used in the side walls of the apse and in one wall of the north wing of the transept (Figure 3.5). This solution satisfied the need to contain the extent of the excavations, which would have been greater with continuous trenches; at the same time, the depth of the localized excavation was better suited to the irregularities of the slope. This is confirmed by the same differences between the foundations of the two sides of the apse: on the south wall there are two arches at different levels, whereas on the north side, there is just one arch with a wider span, which is positioned for the most part at the same level as the analogous foundation arch that can be seen in the adjacent wall of the transept. On the north side, we can see the masonry structure of these arches, with the intrados, orientation and height of the very irregular ashlars; this would suggest that the arches between the pit foundations were built on earth centring and under working conditions that did not support the careful production and laying of ashlars.

In view of the distinct geological features of the site, the levels chosen by the builders of the new cathedral are significant: a room was created under the apse and its south wall had a foundation wall offset, which could roughly indicate a walking surface, placed just 1 m below the offset of the apse of the old cathedral. Before this space could be entered from inside the cathedral it was accessed through three doors created at the base of the back wall of the apse on Vallepiatta (Figure 3.6). With respect to the wall offset, the threshold of the three entrances was lower by an additional 60 cm; this meant that the excavation area in the crypt potentially did not exceed a depth of 150 cm.

Continuing with the examination of the specific wall elements, the focus now shifts to the remains of the apse and the transept visible in the spaces below the cathedral. The apsidal external facing of the façade, with its three entrances, can be identified and inspected in correspondence to the central door, now above the apse of the baptistery. It is made of carefully squared cavernous limestone ashlars worked with a toothed hammer and given a framing band, laid with thin joints and corner colonettes in the jambs of the portal. Moreover, the base of the wall is made up of a plinth surmounted by moulding (Figure 3.7). The façade was delimited at the sides by corner pilasters; the one to the north is partly visible. The side walls of the apse attach on the south to a section of one side of the transept, and on the north to a whole side and part of the back wall of the other wing of the transept. These external walls are also made of limestone, but they have rather roughly squared ashlars placed in

Figure 3.5 The arches of the pit foundations emerge in the sides of the crypt.

continuous rows. On the south wall of the apse and that adjacent to the transept the limestone masonry rises almost as far as the current floor of the church, and lower down the foundations revealed during excavations are visible. In the northern wing of the transept, the top part of the external wall on the east has an uneven brick area. It should be taken into account that these wall sections remain from when the transept and apse were torn down; during the demolition, some parts of the walls may have been excessively reduced and later brought back to the necessary height with partial reconstruction. The external walls of the transept-apse must have therefore been entirely made of limestone and plastered, apart from the rear façade.

Inside the apse wall perimeter, above the footing in the space of the present-day painted crypt, the lack of plasterwork on some parts of the walls allows us to see the brickwork. The

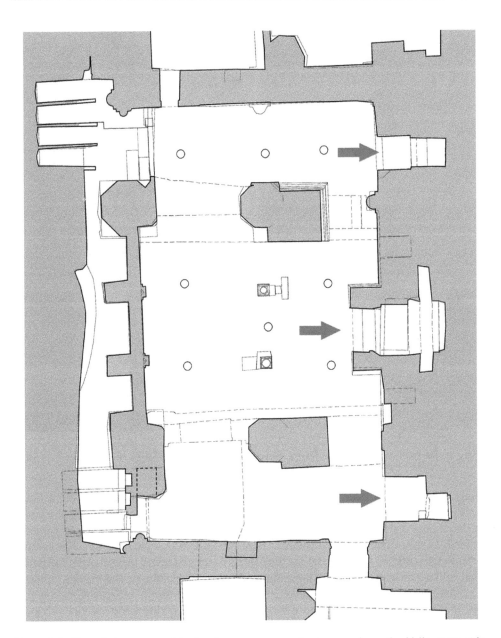

Figure 3.6 Plan of the crypt with the three entrances (red arrows) on the Vallepiatta side.

precision of the masonry and the surfaces of the bricks marked with lines suggest that they were prepared for the laying of painted plaster.

In the room below the apse, with its three entrances from the outside, additional important finds were discovered, attributed to the cathedral they had started building. These finds include semi-columns with side pilasters placed in the corners between the apse and the

Figure 3.7 (a) Base of the façade facing Vallepiatta. (b) Detail of the semi-columns in the corner between the crypt and the transept.

transept, made of brick like the internal walls and given travertine ionic capitals and Attic base (Figure 3.7). The implications of such a complex architectural element near the transept, in an area that remained buried, indicates the intention, which was then discarded, to extend the space that had been created below the apse in this direction too. The corner pilaster to the north, on the side near the transept, is decorated above the balteus of the ionic capital with a human protome with large, rimmed eyes, a wide flat nose and big lips, reflecting a construction phase still far from the arrival of the Pisano family in Siena (Giorgi et al., 2009, 15, 18). We shall return to this room later, which was perhaps also used as a church and entrance hall to the cathedral (Seidel, 2003, 37–41, 74–76) and reused as a crypt, as it conserves traces of many changes made to the cathedral, including some important construction implications.

3.1.3 The basilican body

The initial apse-transept was not immediately joined to the basilican body that it should have corresponded to, but instead it was interrupted outside the old cathedral. This choice might have been dictated by the intention to ensure continuity of worship using the apse area of the earlier cathedral for this purpose, while the aisles of the latter must have been implicated in the construction work for the new cathedral. For the same reason, the underground room with three entrance doors was not initially extended to the transept, but instead it was limited to the rear area of the apse of the old cathedral (Figure 3.8).

The elements that created a break between the initial part of the apse and the transept and the subsequent basilican body are important. The span of the apse opening, approximately 20.30 m, could have implied a nave of the new cathedral so wide that the idea was abandoned. The total width of the basilican body, approximately 24.65 m, exceeds that of the initial apse and it was divided into three aisles. With this dimension – the distance between the pillars of the nave is approximately 10.40 m – the basilican body could be vaulted. This solution was perhaps considered more advanced than an aisle spanning 20 m like the apse, which would have required a trussed roof. The aisles, covered by groin vaults made of bricks roughly 30 cm thick, therefore one-brick thick, were distinct as they did not use ties and extremely small external pillars (roughly 70 cm wide and 60 cm deep) placed up against the clerestory and resting on the diaphragm arches interposed between the bays of the aisles. The intention was to avoid using buttress walls and flying buttresses (Figure 3.9).

There was also a profound change in the choice of materials with respect to the initial part: in the visible facings, two colours of marble, black and white, were used (Figure 3.10), documented from 1227–1229 (Giorgi & Moscadelli, 2005, 89; Causarano, 2009, 207). Note that bricks were used in the highest parts of the sides of the external walls of the aisles, where the alternating bands of black and white were obtained by using painted plaster. A similar solution is found in the lower part of the external side of the clerestory, now incorporated into the garrets: here the duotone stone is interrupted by a travertine cornice and a plastered brick wall continues below. Part of it can be seen in the garret of the north aisle where there is brick masonry section with a fragment of plaster painted in black and white bands, similar to the aisles.

With respect to the basilican body, the position of the bell tower, made of limestone ashlars up to the level of the garret, is indicative, as it became partially incorporated into the south

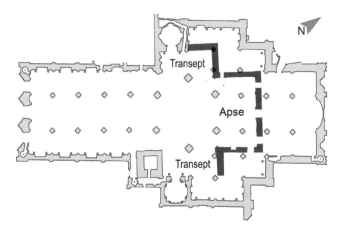

Figure 3.8 Plan of the cathedral with the remains of the apse and transept of the intermediate cathedral (brown).

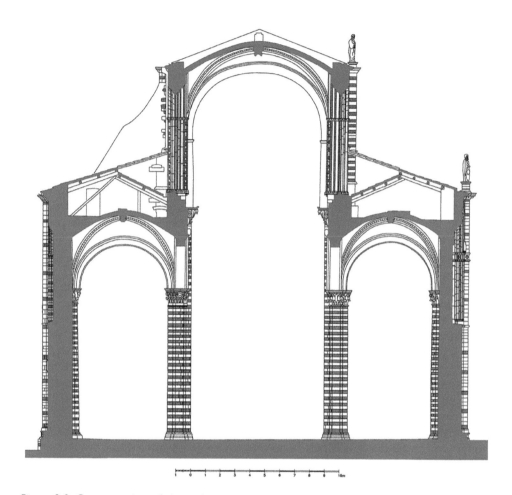

Figure 3.9 Cross section of the aisles.

aisle. It is also clearly rotated with respect to the cathedral they had started building; it is plausible that it was part of the original cathedral complex (Tragbar, 1997, 159–186). Therefore, assuming that position of the bell tower was consistent and parallel with the original cathedral, this would indicate that the naves of the church under construction were rotated with respect to the original one and that the walls were in different positions. In other words, the masonry or foundations of the longitudinal walls of the earlier cathedral do not seem to have been reused, which must have had narrower naves than the new one they had started building.

The limestone section of the bell tower is surmounted on the east by a brick Lombard band that decreases with the same inclination as the roof slopes, while on the north side, there is an overhanging cornice: in both cases, perhaps the idea was to create a support for the wooden beams of the sloping roofs.

The brick part and the raised part of the bell tower created with duotone stone material similar to that used in the basilican body should therefore be attributed to the construction phase that introduced this material. Moreover, following the criterion employed in the aisles

Figure 3.10 View of the nave from the drum.

to use material considered less valuable in the parts that were barely visible, the area below the first window of the bell tower was constructed with bricks on just the side facing the cathedral screened by the volume emerging from the nave and the drum.

Despite the strongly uniform appearance of the duotone basilican body, here too, there are signs of different building phases and consequent unique construction features.

The three central bays of the present-day basilican body were completed with a façade that was set back with respect to the existing one. Traces of it remain in the first two pillars to the west of the present-day church. They are cruciform pillars made up of semi-columns in the two longitudinal and transverse axes and a slender two-sided dihedron in each of the four corners. On the edge of the transverse semi-columns, on the side facing the façade, both pillars show clear masonry repairs. This is thought to be masonry evidence of the existence of a first façade, which had perhaps reused part of the façade belonging to the original cathedral, against which stood pillars, in this case with three lobes instead of four (Haas & von Winterfeld, 1999a, figs. 345–347, 350–351; Haas & von Winterfeld, 2006a, 395–396; Quast, 2007, 105–110). The attachment with the façade created with over half a pillar, less common than the semi-pillar, can also be seen in the bay adjacent to the façade of Arezzo cathedral.

3.1.4 The hexagonal crossing

As the works progressed from the basilican body to the crossing, the pillars west of the dome were put in place, re-proposing the four-lobed solution oriented on the longitudinal and transverse axes; but the section of the pillars was increased by adding diagonal bands and keeping the semicircular lobes the same size.

This led to the completion of the first structure of the basilican body, formed of four bays in this phase. The last bay built is as wide as the remaining ones on plan, except for the unique features created by the presence of the bell tower, which is partly incorporated into an aisle. However, the completion span has distinct features which are worth considering.

The groin vaults of the aisles have a somewhat rampant and slightly curved generatrix, while in the vaults of the aisles constructed earlier, the generatrices are much more rampant and have a more accentuated curvature. It would seem that close to the pillars of the dome, the aim was to create a more rigid vaulted system (Haas & von Winterfeld, 1999b, tavv. 14–15) (Figure 3.11).

A further aspect concerns the bay of the nave, where the magnificence of the triumphal arch that leads to the domed area should be pointed out; it features concentric duotone mouldings of serpentine and marble and has an overhang created by numerous and varied profiles with alternating bands of leaves, eggs and dentils. The keystone area is embellished with a lion head wedged into the intermediate band of egg and dart moulding and dentils, and the head of a young man inserted into the extrados band. The top of this arch is at the same level as the cornice, in the domed space, supporting the blind loggia inside the drum, where there are corbels with numerous busts. Further similar refined motives can be seen in the capitals of the northeast trifora (three-light window) of the drum. These motifs had "rare expressive power" attributed to Nicola Pisano (Bagnoli, 2003, 148–151), whose presence at the construction site was documented in 1259 (Giorgi & Moscadelli, 2003, 90; Giorgi & Moscadelli, 2005, 71–72). The set of these elements and their particular location and diffusion suggest that Nicola Pisano was involved in the architectural definition of the domed space at least up to the base of the blind loggia of the drum (Figure 3.12).

The layout of the dome pillars, which form a hexagon, is distinct in terms of their sizes and positioning, which is worth focusing on. The side with two pillars on the west is clearly rotated with respect to that of the pillars on the east side. In fact, the line joining the centre-to-centre distances of the pillars on the west compared to that of the pillars on the east shows a rotation of around 3.5° between the pairs of pillars. This considerable rotation stems from the fact that the pillars on the west are aligned with the bays of the nave, while on the opposite side, the pillars are placed at the opening of the apse constructed in limestone. This means that thrust of the arches, through the diaphragm arches of the aisles, would have been offset on the east by the transept walls, on the west on one side by the bell tower, on the other to the north by a transept wall, on which one of the major sides of the Piccolomini Library was later built. It is no coincidence that the two pillars on the west are rotated with respect to this wall: this means that the basilican body is rotated with respect to the apse and the transept.

The orientation of the north and south pillars of the hexagon is borrowed from that of the east and west ones, assuming an intermediate rotation to them. However, the position of the north and south pillars creates a hexagon, leaving aside the general irregularity of the sides, with the oblique sides to the east being much shorter than those on the west: the centre-to-centre distances between the pillars are, respectively, 9.37 m (northeast) and 9.70 m (south-east), 10.53 m (northwest) and 11.14 m (southwest). It is not known if such a clear shift

Figure 3.11 Cross section of the aisles of the bay adjacent to the crossing.

to the east of these pillars is due to fractures in the ground below, identified by geological surveys, or whether it was determined by the architectural structures of the earlier cathedral. This latter, following the construction of the pillars of the crossing, can now be considered definitively dismantled also in the apse area. The use of the ancient cathedral is documented at least until 1220 (Butzek, 2010, 111–112).

   With respect to the irregular dimensions of the hexagon of the crossing, there is distinct architectural congruity between the pillars on the east and west sides and the corners of the hexagon starting from the springer of the arches. These corners are shifted by the axes of

Figure 3.12 Details of the drum, trifora (a) and cornice with corbels at the base of the blind loggia (b).

the pillars so that they support the elevation of the semi-columns inside the hexagon beyond the springer of the arches. In the north and south pillars only, the semi-columns continue beyond the springers of the arches without flanking pilasters. This clever architectural device connects the lobed pillars, oriented with longitudinal and transversal axes, to the hexagonal drum and conveys the intention to transform the squinch pendentives, located at the corners of the hexagon, into niches housing sculptural works positioned on the semi-columns rising from the pillars (the statues of saints now placed on these semi-columns date to 1488 and were installed in 1490) (Carli, 1979, 109–110). This is a unique solution that, with the objective of implementing the sculptural values of architecture, suggests that Nicola Pisano may have been involved in its conception. The lobed pillars with a semi-column flanked by side pilasters that continue between the springers of two arches were however also placed in the nave of the basilican body, where these elements rise up to the cornice at the base of the clerestory. The innovative element of the crossing, with respect to the basilican body, was to combine it with semi-columns that continue up to the squinches of the pendentives.

The pillars of the dome connected three different sections of the cathedral: the initial apse and transept, the basilican body and the bell tower of the earlier cathedral. In a similar context, the construction situation of the arches that stand on the north and south pillars was quite distinct: since the pillars were placed roughly halfway up the arms of the transept, here the thrust of these arches, unlike that transmitted to the arches on the east and west sides, could not be offset by a wall achieved by means of a diaphragm arch; in fact, a system of branched diaphragm arches that reached different points of the perimeter walls of the transept had to be used. It should be said that the arches on the skewed sides of the hexagon are not as high, but they support the greater masonry masses of the drum. The specific static condition of the north and south pillars led to them being constructed, compared to the others, with a resistant section about 30% larger (the section of the east and west ones is roughly 2.94 m²; that of the north and south ones 3.70 m²). The checks carried out revealed that these pillars have remained vertical over time.

It is worth mentioning a size comparison between the crossings of the cathedrals of Siena and Pisa. In Siena cathedral, considering the centre-to-centre distances of the pillars at the entrance to the apse and the start of the nave, corresponding to the width of the transept (Haas & von Winterfeld, 1999b, tav. 3; Giorgi et al., 2009, 7), the longitudinal dimensions are 17.30 m (north) and 17.96 m (south) and the transversal dimensions are 10.20 m (west) and 10.65 m (east). In Pisa cathedral, the longitudinal centre-to-centre distance is 18.20 m and the transverse one 13.50 m. The longitudinal dimension of the two crossings coincides for the most part, while the width of the bay in Siena cathedral is significantly smaller. In this case, to find the space necessary to construct the dome, the intermediate supports had to be moved laterally, beyond the width of the nave.

3.1.5 The drum and the dome

The phase in which the architectural solution of the crossing was connected to the bay of the basilican body produced some unique construction features. The overhang of the triumphal arch led to an adaptation at the level of the clerestory of the adjacent bay, where in order to place walls of a similar width to the triforas the latter had to be shifted from the axis of the bay. This repositioning resulted in walls of different widths flanking the triforas. As a result, the blind arches placed to the side of the triforas on the outside have clearly different spans here (2.42 m near the drum, 1.79 m on the opposite side).

The remaining triforas of the clerestory of the basilican body, still divided into four bays today, are aligned with the bays and flanked by small blind arches of the same span. These blind arches are not merely decorative, as the masonry above them has an overhang equivalent to their thickness. In other words, this created an external wall offset which was concealed by the same arches. A similar solution can be seen in the wall of the attic of Florence baptistery, where on the inside the wall was thickened by means of an overhang placed on a quarter-circle moulding (Rocchi Coopmans de Yoldi ed., 1996).

At the level of the loggias of the drum there are several traces of a subsequent different construction phase. In fact, at the end of the nave a transverse arch was set, or constructed, in the position of the current paired columns and it was therefore incompatible with them. The remains of this arch were partly chiselled to make room for the present-day columns, which were necessarily planned after this arch (Figure 3.13).

The connection between the sides of the clerestory of the basilican body and the paired columns of the drum was made in a single construction phase. On the side of the nave

Figure 3.13 In the connection between the nave and the drum, the remains of the springer of an arch can be seen (red arrows), which is incompatible with the double columns of the loggia.

towards the façade, the columns are placed at regular intervals. They are interrupted for the section of the clerestory walls and continue with continuity around the remaining perimeter of the drum, all placed on the outside (as we will see later, on the east side some of the columns were subsequently incorporated by the sides of the nave). In this phase, there was an apse with a roof placed at a lower level than the paired columns of the drum. This was possible due to the characteristics of the arch of the crossing in front of the apse which, with respect to the one opposite near the nave, not only had simplified moulding, but it was also lower. This was how the space in which to place the roof of the apse between this arch and the crown of columns above was arranged (Figure 3.14).

The narrow corridors created between the end walls of the clerestory and the drum were covered by a flat stone lacunar ceiling with coffered panels containing a flower. This refined ceiling was also adapted to the different widths of the corridors, and the size of the lacunars was brought into line with them (Figure 3.15).

This part of the drum is also duotone, white and green stone, but in this case, each band is made up of two rows of ashlars instead of one, as in the lower levels of the drum and the basilican body. The same wall structure solution was used in the raised part of the bell tower; by 1259, work was in progress; and it had to be completed in 1264 as soon as the roof was finished (Giorgi & Moscadelli, 2005, 100 note 89). The duotone double row is an additional unique feature that distinguishes the building phase that started from the level of the drum loggias.

It has been hypothesized that the paired-column loggia of the drum, at an early stage, was visible around the entire perimeter and that it coexisted with a vaulted basilican body that was lower than the existing one. This hypothetical arrangement, which is not in any way confirmed by the concrete elements examined above, would be proven by the external blind

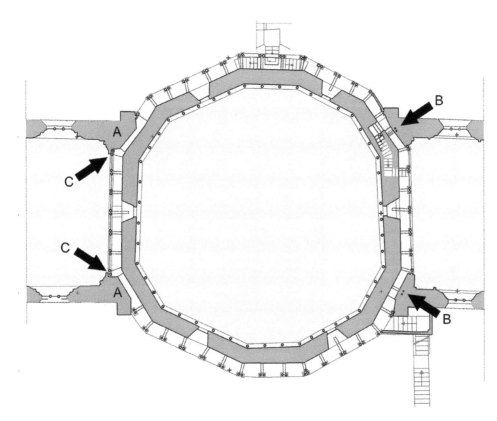

Figure 3.14 Plan of the drum at the level of the lower loggia; on the side of the aisle the masonry (A) is coeval with the loggia; on the side of the apse, the paired columns are incorporated into the masonry (B); position of the arch springers (C) which are incompatible with the columns of the drum.

arches, considered the original crowning of the clerestory. However, to check this hypothesis, we need only take into account the cross-section of the basilican body and the longitudinal section of the south aisle with a view of the clerestory wall (Haas & von Winterfeld, 1999b, tavv. 15, 21) in order to ascertain that at such a height, the roof and the vaults of the nave would have largely overlapped the magnificent triumphal arch that opens onto the crossing, making this hypothesis unfounded even for this reason alone.

In the crossing the two largest arches are on the longitudinal axis, while the considerably lower height of the arches on the skewed sides to the north and south places the start of the drum on these parts at a lower height. So the two larger arches are flanked by triforas in each of the skewed sides.

The irregularities of the basic hexagon are reflected in the dimensions of the triforas; those on the larger sides to the west have wider arches and larger pillars (Figure 3.16). These latter have a section of 0.75 m² on the west sides and 0.60 m² on the east sides. It should be pointed out that with respect to the variability of the sides of the crossing, these size adaptations were made with great precision, as evidenced by the existing dimensional homogeneity, between the larger triforas on the west and the smaller ones on the east. All the triforas on the outside

Figure 3.15 The narrow corridors between the walls of the nave and the drum are covered by a coffered stone ceiling.

are placed in a slight recess topped by an arch; the recess is deepest on the east sides, precisely where the pillars reduce in size. Another unique feature is found in the trifora on the northeast side. Here, in the archways of the external side, there are splays in the side walls that are deeper than those of the pillars. Moreover, in the initial part of the masonry, there is a single splay which did not provide for the trifora to be set back with respect to the external arch. There are traces of changes made during the works: initially, there was no trifora set further back than the drum wall; then a splay was created inside an arch, but moving on to the construction of the trifora, it was decided to reduce the dimensions of the two intermediate pillars, which therefore had splays that were not as deep as those of the piers of the arch. In the remaining cases, in fact, the splays of the pillars have the same depth as those of the side walls (Figure 3.17).

Despite their skill, the builders could not avoid irregularities in the layout of the crossing which, as seen, can be traced back to the project concept, defined by a gradual process and variations, and influences brought about by the context. Other unique but less evident features were created as many construction details were defined during construction.

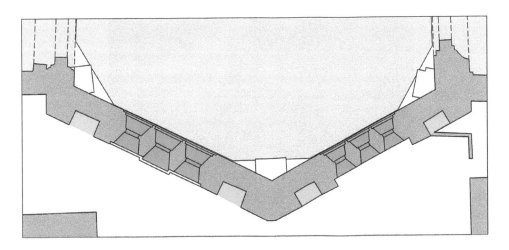

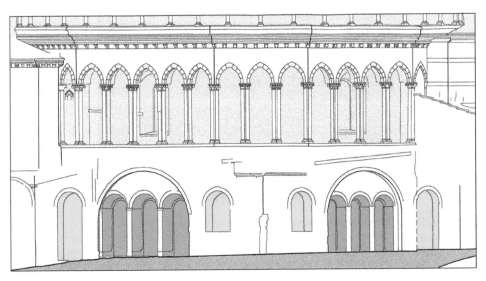

Figure 3.16 Excerpt of the surveys of the hexagonal drum on the bell tower side; the triforas (green) have different widths; the shorter niches (light green) are positioned in the corner of the hexagon.

Great attention was also paid to the execution details of the walls. Evidence of this can be seen in the duotone areas found in the splays of the arches of the triforas and in the squinches of the pendentives, made up of precise wedge-shaped ashlars that outline the semi-cone shape of these structures. This geometric definition is an approximation in some cases, as the pendentives are sometimes enlarged in the innermost area. To date, it has not been possible to examine this aspect with the necessary care; however, it could be a modification that was made when the statues were installed in order to obtain more space inside the pendentive. In order to avoid the difficulty of connecting the horizontal duotone bands of the walls with the radial duotone parts of the arches and pendentives it was decided, most likely during the works, to

Figure 3.17 Detail of the trifora on the north side of the drum; the width of the initial part of the recess (green) was then reduced to create a shallower recess flanked by a pillar where the arch above rests (cyan); thereafter, in the intermediate pillars, the width of the recesses and of the arches above was further reduced (yellow).

interrupt the continuity of the walls between these elements by inserting a cornice between them; moreover, the archivolts are slightly set back with respect to the adjacent facings. In the case of the triforas, this cornice on the edge of the archivolt rests on small columns, while in the pendentives, it is placed on corbels. On the external side of the drum, it was decided not to create a cornice framing the archivolts of the triforas; here the lack of continuity between the duotone wall and the duotone archivolts is evident. But it is likely that they also wanted to embellish the architectural decoration in the interior.

Given the skill of the builders, it may seem strange for the pendentives near the nave to partly overlap the triumphal arch, determining the lack of a corbel and, for a stretch, the arch above on the edge of the archivolt. This aspect can be explained if we observe the relationship between the arches of the triforas and the pendentives, paying attention to the parts where these elements touch. Here, the arches have a stretch of interrupted cornice on the external support and a similar horizontal cornice begins, a little higher up, above the corbel of the pendentive. It would seem that the initial idea was to create a continuous cornice that continued from the springer of the trifora arch up to the springer of the pendentive. However, when constructing the pendentives, it was probably decided to raise them by about two rows precisely in order to reduce the overlap with the triumphal arch in the east pendentives. This problem did not occur on the opposite side near the apse, where the transverse arch is lower in height. On the other hand, the reciprocal position between the inner corner of the penden-tive and the triumphal arch was established from the top of the pillars of the crossing, where the corners of the hexagonal part of the drum are positioned, which dictate the position of the pendentives and the springer of the archivolt of the arch. The partial overlap between the triumphal arch and the adjacent pendentives is one of the many construction adaptations that were inevitable in architecture of such size and complex origins.

Moreover, the proximity to the triumphal arch may have limited the size of the penden-tives themselves, which are also very close to the triumphal arches. The pendentives mark the shift from the hexagonal plan to the dodecagonal plan of the drum, creating internal sides of 3.35/3.60 m, with respect to the larger sides of 6.70/7.15 m. So the shift from the hexago-nal plan to the dodecagonal also accentuates the geometric irregularity of the plan. The mod-erate depth of the pendentives (1.00/1.5 m) is equivalent to the amount the walls above them were repositioned towards the inside of the crossing. These walls were repositioned towards the crossing, so in the outer corners of the hexagonal drum there are wall discontinuities with a triangular plan similar to that of the pendentives.

The diagonal sections of the crossing that intercept the central openings of the trifora high-light an important construction aspect of the drum. In the meantime, it should be pointed out that on the diagonal sides of the crossing, there are triangular bays covered by two webs. The arrangement of this vaulted system, which as we will see can be attributed to a later building phase, is equivalent to half a groin vault on a quadrilateral plan: the arch positioned on the side of the crossing constitutes the diagonal rib of a hypothetical vaulted system on a quad-rilateral plan. At the keystone of the diagonal arch, we find the other diagonal rib interposed between the two webs.

The diagonal sections show that the webs of the triangular bays extend for more than half of the wall of the triforas (Figure 3.18). However, it seems unlikely that the pillars of the triforas and the adjacent walls rest directly on the webs below. There is probably an arch between these structures. On the lower side of the openings of the triforas, there is a rustic splay (with putlog holes testifying its execution was coeval to the trifora) that rises by means of offsets in the direction of the external side: this would have provided sufficient thickness to create an additional arch above the webs. In fact, two distinct arches must have been cre-ated below the triforas, and the one near the crossing may have had a thinner section.

In any case it is unlikely that the loads from the drum-dome system were transferred directly to the triforas. The slightly overhanging arch on the external side, just above the triforas, could suggest the existence of a relieving arch extending for the entire wall thick-ness. These would be relieving arches, above each trifora, placed at a similar level to that of the two arches of the crossing transversal to the nave. The construction system is made even more complex by the niches flanked on the external side by the triforas and almost as deep as half the wall thickness (Figure 3.16). In fact, relieving arches should be hypothesized placed in the section of masonry between the triforas and niches, or in a position corresponding to the haunches of the underlying arches of the crossing. The masonry adhering to the haunches of these arches actually creates a system of diaphragm arches on which the drum stands.

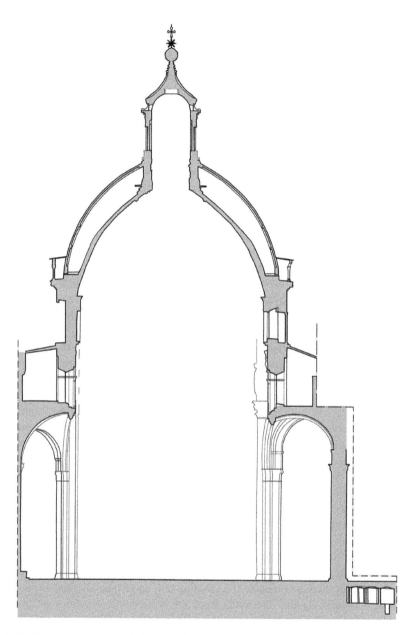

Figure 3.18 Diagonal section of the crossing in the centre of the north and south sides.

Concealed arches, like those placed under the triforas of the drum, are also found in the longitudinal arches of the nave of the basilican body (as well as in the presbytery, built in a later construction phase). The longitudinal arches, as documented by the cross-section, are considerably narrower in width than the wall of the clerestory above. This creates a wall that extends to the webs of the north and south aisles up to a width of 40 cm (Figure 3.19).

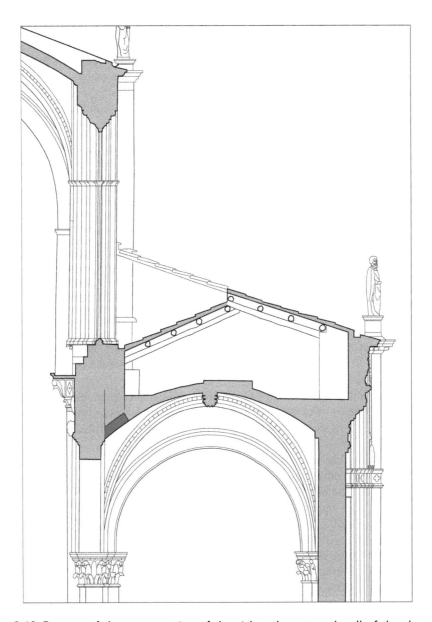

Figure 3.19 Excerpt of the cross section of the aisles; the external wall of the clerestory seems to overhang (red) a section of the groin vault.

The explanation for a similar arrangement could come from the transverse arch of the crossing facing the presbytery, which has a very deep splay on the apse side. A similar solution could be hypothesized in the arches flanking the aisles, with the webs placed in continuity with the splays. This created bays covered by groin vaults that are apparently wider than the actual distance between the wall of the clerestory and the outer sides of the aisles.

The shorter niches flanking the triforas are placed close to the north and south corners of the hexagonal drum; in some cases it has been possible to confirm that brick masonry was used at the base and the edge of these niches; while the taller niches are placed on the opposite side of the triforas and have a duotone stone facing around the entire perimeter. The size of the niches and the arrangement of the different materials, stone ashlars and bricks, suggest the possible position of the two-pitched roof of the transept, with the roof ridge roughly corresponding to the north and south corners of the hexagonal part of the drum (Haas & von Winterfeld, 2006a, 421–427, figs. 42–43).

In the north corner of the hexagonal drum, in correspondence to the pendentive, an initial short stretch of stairs was built, which continues to climb with two further ramps going in opposite directions, created in the adjacent sides of the drum, until reaching the level of the loggia with paired columns (Figure 3.20). Between the first stretch of stairs and the pendentive, the wall thickness is greatly reduced (by about 25 cm); consequently, access to a loggia of the drum was created while the pendentives were being built. It should be noted that the first stretch of stairs used the external space created in the corners of the hexagonal drum, passing through the pendentives to a dodecagonal drum, while the two continuing stretches of stairs were created in the sections of the hexagonal drum below the loggia.

The initial phase of the drum was concluded with the creation of the pendentives and the larger arches of the crossing; it was followed by the decision to create a section of the drum with a double loggia. An internal blind one, with small columns adhering to the masonry and surmounted by an entablature, and an external one with paired columns and a walkway. As regards the external loggia in particular, it has already been pointed out that it is a variant because on the side of the nave of the basilican body, an arch that was not compatible with the paired columns was planned.

As regards the vertical continuity of the structures, the blind loggia and the lintel above overhang the walls below. The stability of the small columns has been reinforced here with metal hooping around the shaft, connected by a dowel to the masonry behind which is placed at a distance of about 5 cm.

The paired marble columns of the external loggia sit on the edge of the masonry below and are remarkably slender: they have a diameter of 15 cm and a height of about 2.74 m (including the bases and capitals) (Figure 3.21). The loggia is flanked on the inside by a dodecagonal wall with very irregular sides which vary from a minimum length of 3.80 m to a maximum length of 7.60 m. In the masonry of the drum, from the hexagonal part to the level of the loggias, the thickness reduces considerably, from 1.70 to about 1.00 m; it must be said, however, that the masonry is more continuous at the level of the loggias, the openings here are limited to six arched windows placed on the longitudinal axis of the cathedral and on the sides above the triforas. In addition, the wall mass was reduced due to the staircase that leads up to the base of the dome.

The width of the bays of the loggias, apparently homogeneous, was actually adapted to each side of the polygon. In the blind loggia there are sides with column centre-to-centre distances of around 1 m and sides in which this measurement extends to 1.30 m. In the external loggia, the width of the bays varies from around 1.02 to 1.22 m.

The most interesting construction aspects are in the loggia with paired columns. These latter support a transversal stone lintel which, on the inside, is usually inserted into the masonry. This solution in fact posed particular problems at the arched windows as here the lintel is positioned in the space of the openings. Two solutions were adopted in these cases: whenever possible, the transverse lintels were positioned at the edge of the arched

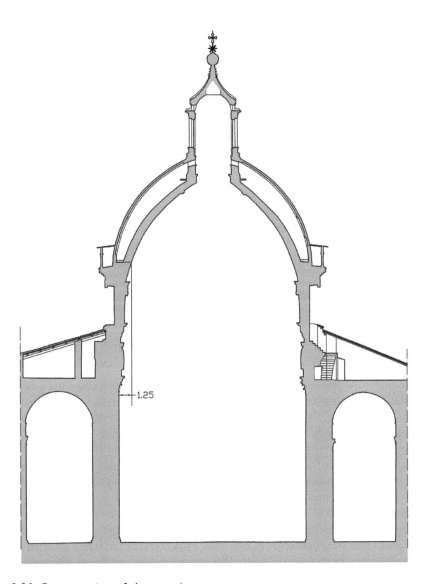

Figure 3.20 Cross section of the crossing.

windows, giving them a skewed position. When it was inevitable that the window opening would be intercepted, as is the case with four openings, the transverse lintel was positioned in the centre of the opening, also using slightly skewed positions, where a joint creates a connection to a vertical stone element which, in turn, rises upwards to fit into the arch of the window. So the stone lintel is suspended on a vertical stone element which at the same time acts as the keystone of the arch.

This bold solution seems necessary considering that the openings are for the most part placed in the centre of the respective sides and that on the inner side they are positioned between the intervals of the columns of the blind loggia. The difficulty of creating a

Figure 3.21 The dome and the two levels of loggias; (left) the central part of the cathedral's bell tower.

coherent architectural and constructive relationship between the openings, the inner loggia and the outer loggia is also evident on the north-east side of the latter, where the columns flanking the window are placed at a much larger distance than on the remaining sides in order to avoid placing a small column in front of the actual opening.

Brick pointed barrel vaults rest on the transversal lintels. It is not surprising that the construction system formed of stone lintels connected to vertical stone elements needed to be reinforced with metal brackets; in the north-east opening, an L-shaped bracket was inserted to support the stone elements. In the east window, the vertical stone element has been replaced by a metal bracket; the date 1836 painted in the bracket area most likely refers to this intervention (Figure 3.22). The horizontal and vertical stone elements still appear intact in the south-east and west openings. In any case, even some of the lintels inserted directly into the masonry have cracks, and it is clear that they have been reinforced with metal bands.

The transition area between the first level of the external and internal loggias and the base of the dome is of great interest due to the constructive interconnections between apparently distinct elements (Figure 3.23).

The wall adjacent to the internal loggia is tilted slightly inwards; this is highlighted by the diagonal sections and the columns of the loggia, which are visibly more detached from the wall at the bottom than at the top, where in some cases, the annulet of the capital is tangent to the wall. The slight tilt of the wall may have been achieved with overhanging rows as preparation for the support of the dome. In fact, if we imagine these walls rising vertically, the support of the dome positioned on the entablature of the columns would have had a much greater overhang than this wall. From the sections, it would seem that the aim was to achieve continuity between the curve of the dome and the tilt of the wall at the level of the internal loggia. The considerable size of the entablature crowing the columns of the loggia should

Figure 3.22 Paired-column loggia of the drum: details of the stone lintels that connect the columns to the interior wall of the drum; many of the lintels are cracked and strengthened with metal bars using different solutions.

Figure 3.23 Blind loggia of the drum and view of the nave.

be pointed out: at around 1.27 m height it seems more like a support for the dome than the entablature of columns approximately 2.58 m high. Prophets and patriarchs of the Old Testament are depicted in each bay between the columns of the blind loggia. One of them, low down, bears the signature of the painter, "OPUS BENVENUTI" (Figure 3.24). It refers to Benvenuto di Giovanni, who was paid on 19 August 1482 for having painted some of these figures (Carli, 1979, 119–120).

The work to prepare the support of the dome, which started with the construction of the loggias, is even more evident at the height of the springer of the dome. This level corresponds to the band of lozenges running along the arches of the external loggia, a much lower level than the crowning cornice of this part. The wall was therefore built above the brick vaults of the loggia, which were half-a-brick thick, at the same time as the initial part of the dome. This single masonry unit rises for a further stretch in the area closest to the base of the dome, placed at the edge of an additional external loggia. The tests carried out on the walkway of this loggia, which flanks the entire perimeter of the dome, highlighted a brick wall most likely extending from the intrados of the dome to the outer facing in duotone white and green stone, topped off by corbels, reminiscent of those below the internal blind loggia, and a prominent marble cornice. The aim was to create a single wall unit that, partly acting as a corbel due to the quality of the mortar, may have also helped to limit the loads weighing directly on the vaults of the external loggia.

The start of construction of the dome is linked to the creation of all the loggias. The dome has a shell structure approximately 1.85 m above the entablature of the internal blind loggia, where it starts to rise with masonry one-and-a-half-brick thick.

The construction of a masonry connection between the loggias and the base of the dome created the conditions for the start of the autonomous structure of the dome to be placed on a

Figure 3.24 Detail of the pictorial decoration alternating with the columns of the blind log-gia of the drum; the base of an image of the cycle of Prophets and Patriarchs shows the signature of the painter, "Opus Benvenuti."

sort of wall corbel which, with respect to the pillars of the crossing below, had an overhang varying from 1.25 to 1.40 m (Figure 3.20).

The tests carried out on the extrados of the dome, covered with plaster, revealed bricks placed in horizontal overhanging rows up to at least 2.5 m above the walkway positioned at the base of the hollow space of the dome. This construction technique was most like used in continuity starting from the tilted walls behind the blind loggia of the drum. This masonry structure corresponds to the laying technique already seen in the dome of the Baptistery of Florence (Giorgi & Matracchi, 2017, 193–207) and in that of Pisa cathedral. If we consider the vertical rise of the dome of Siena cathedral, from the springer moulding to the summit, it is clear that almost half its height was built with horizontal overhanging rows.

The base of the dome aligns with the irregular dodecagon created by the pendentives. It is made up of twelve webs with different extensions and curvature; there are webs with a continuous curved profile as well as webs whose profile is broken into two/three almost straight segments. The construction was likely completed in the section with horizontal rows with tracing at the corners, while actual centring was used in the remaining top part. The laser scanner survey of the intrados produced horizontal contour lines that revealed the gradual process of regularizing the dome, which in the highest part eliminates the distinction between the webs by creating a hemispherical cap to a dome with a circular plan (Figure 3.25).

The changes in the curvature of the dome sometimes coincide with the change in thickness, which goes from masonry of one-brick-and-a-half to one-brick thick, creating a wall offset at the extrados.

The bricks are quite regular, at least in the areas where they could be checked: they are 5.4/6 cm high, 13.2/13.5 cm wide and 29/31 cm long; they are laid on beds of mortar with a thickness of 1.5/2 cm. At the bottom in particular, however, there are strong variations

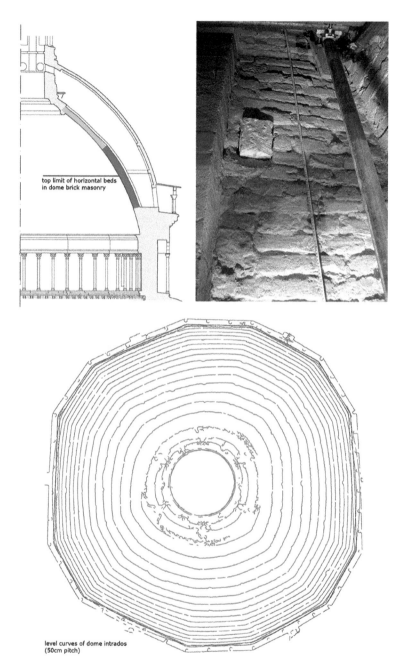

Figure 3.25 (c) Contour lines of the intrados of the dome obtained from a laser scanner survey; in the top part (red), the corners of the dodecagonal plan gradually disappear. (a) The section highlights the different thickness of the medieval dome, one-and-a-half bricks in the first section (red), one brick in the remaining part (yellow); up to the level of the horizontal line, it has been ascertained that the masonry of the dome is made up of horizontal rows. (b) Brick facing of the extrados of the internal dome.

ranging from a minimum of 0.2 mm up to 4 cm; in the upper part, the joints of the bricks are more regular. In the tests, most bricks proved to be headers. In the lower part, the tests revealed the widespread chiselling of the bricks, perhaps due to the repairs to the plasterwork on the extrados.

The small loggia at the top of the drum is placed against the external dome originally built in wood; this solution was evidently intended to accentuate the external grandeur of the dome structure. The builders must have considered the size of the internal one in masonry insufficient. The external dome, with parts of the nave roofing, was completely destroyed by a fire that broke out in 1890. Photos documenting the damage show few remains of the loggia adjacent to the dome (Haas & von Winterfeld, 1999a, figs. 97–99), which must have been largely destroyed by the fire and consequent collapses. The present-day one should essentially be considered a reconstruction. Recent restorations of the dome have revealed that the small loggia, with its marble columns with a diameter of 15 cm and height of 80 cm, consists of arches generally formed by a marble monolith, with small areas integrated by thin adhering masonry (Figure 3.26). The marble blocks connect with vertical joints approximately in line with the capitals. The duotone appearance of the apparent arches and the green crowning

Figure 3.26 The loggia adjacent to the dome is formed of arches made of marble blocks.

band was obtained by carving the marble block so that green serpentine plugs could be inserted. This is a decorative loggia, not strictly involved in the overall construction device that links the drum and the dome. The huge blocks that form the arches are most likely joined by pivots placed at the top. The free wall created above the arches is now connected to the external dome by a thin metal structure that supports the roof slabs of the same loggia. The connection between the capitals and overlying blocks, checked with tests and instrumental verifications, was created with brass dowels. An additional metal pin ensures the connection between the columns and bases.

3.1.6 Redefinition of the apse and reuse of the ionic crypt

The construction of the east pillars of the crossing and the two pillars placed in the apse area, of which only the foundations survive, involved the dismantling of a pre-existing crypt. Traces of painted plaster were found in the foundations of the two pillars of the crossing, attesting to the demolition of the Ionic crypt that existed at the time. The foundations of these pillars are probably located in the area bordering the first crypt built, limited to the apse area. These foundations are visible in the space adjacent to the crypt, and their shafts are formed of three distinct sections. The upper part of the northern pillar has an octagonal plan and is made of stone ashlars which are visible for 1.10 m, while the octagonal plan of the southern one appears for a short stretch. Below the octagonal part, the base shaft has a roughly square section, with a first stretch made of square ashlars arranged in precise rows. Further down, the masonry becomes rather chaotic, with heterogeneous material also from demolitions, as evidenced by fragments of plaster and characterized by the widespread presence of river pebbles. The lower part of the foundation seems to have been created within excavations against the ground (in the visible parts of the walls there are no traces that suggest the use of formwork), while the ashlar foundation must have been laid above ground level. The well-laid masonry of the two foundations starts at almost corresponding levels, with a difference of around 30 cm; the foundations set in the dug-out area are, significantly, for the most part at the same level as the base of the Ionic semi-columns, that is, around 1.65 m higher than the floor of the present-day crypt (Figure 3.27).

The crypt was also affected by the foundations of two pillars built inside the apse, which other walls were placed up against at a later time. The two foundations have an irregular octagonal plan. In the lower parts without plaster, the conglomerate of the foundation casting is clearly visible with the widespread presence of river pebbles; the use of stone ashlars can be observed in the higher part. However, at the base of one of the two pillars there is a square plan. From this, we can infer that in this case too, a foundation was made in a square dug-out area, which was later brought out of the ground in order to lower the floor of the crypt; the foundation shaft brought to light must have then been reworked giving it its current octagonal plan.

Further masonry evidence of the lowering of the floor of the crypt, with parts of the foundation visible above ground, can be found in the north and south walls; in the latter, the chiselling of the foundation offset is also evident; on the east side, with the three entrances to the crypt, the original higher level of the side doors and the consequent lowering of the jambs is confirmed.

These alterations significantly changed the constructive structure, in particular with regard to the two foundations to the west on which the pillars of the dome stand, which are thus 4.50 m higher, a 35% extension of the original height. In particular, the foundation shafts created in the pit were brought above ground for a height of 230/260 cm, profoundly changing the

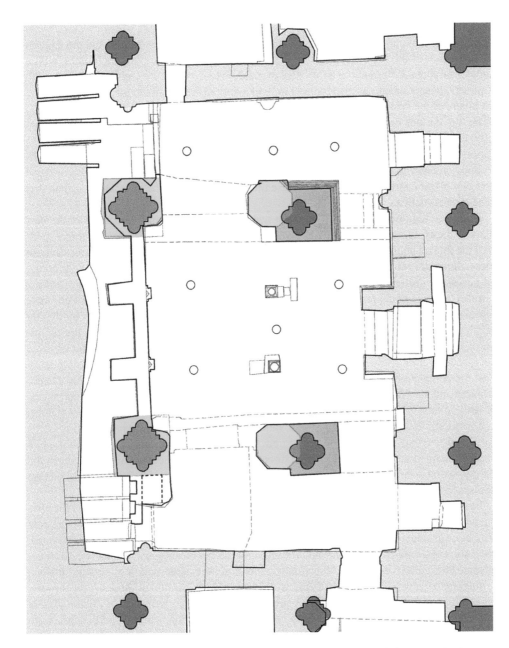

Figure 3.27 The plan highlights the foundations of the dome pillars and those inside the apse of the intermediate cathedral (green); the latter foundations were enlarged (orange) to obtain a support for the pillars of the present-day cathedral.

static conditions. Considering the characteristics of the masonry, all of this would seem to be a risky choice, but the builders probably relied on the quality of the mortar as they knew of its high resistance. If this were the case, there was likely only a short time between the construction of the pit foundations and the excavation of the ground around them in order to lower the level of the crypt.

The most recent archaeological investigations have further excavated an area at the edge of the foundation of the north pillar inside the crypt, where a semi-dug hut was found, with a diameter of about 3.50 m and a depth of 2 m, datable to between the end of the 6th and 7th centuries. These excavations also brought to light a limited section of the pit foundation of the east pillars of the crossing (Causarano et al., 2003, 155–158).

The foundations inside the apse, as they had to bear a lighter load than those of the pillars of the dome, were made with smaller horizontal sections. Moreover, they were placed to support two longitudinal arches of similar width in the apse area. Compared to the basilican body, a slightly wider nave and smaller side spans were built here. The width of these latter was significantly reduced also because the overall width of the apse was less than that of the entire basilican body.

The dismantling of the pre-existing crypt and the levels of the pit foundations of the east pillars of the crossing and of the two inside the apse suggest a building phase that foreshadowed the final abandonment of the crypt. During the construction of the foundations of these pillars, the entire room below the apse, which could be accessed from the three rear doors, was necessarily fully involved in the construction phase and therefore unusable.

In fact, archive documents confirm, in 1255 (1 April) and 1256/1257 (7 and 9 February) (Giorgi & Moscadelli, 2005, 72 note 102), that there were delays and difficulties in making decisions about this part of the cathedral. In order to resolve issues concerning the definition of the apse and the use of the crypt it was necessary to set up a commission as these two aspects were closely linked. A decision had to be made between three possible solutions: the complete filling of the space under the apse, the conservation of the three eastern entrances and therefore the space under the apse, and the partial closure of the rear entrances. So, while one or two of the rear doors of the crypt were actually reopened in 1259 (Giorgi & Moscadelli, 2005, 72–73), at least the vaulted system in front of each of them must be considered as completed at this time. The new layout of the reused ionic crypt, with the two massive intermediate pillars and lowered floor levels, likely included an aisle in front of the side doors and a three-aisle space in front of the central door (Seidel, 2003, 62–74); these three spaces could have been made accessible in different phases.

The complex organization of the construction site also concerned the definition of the apse structures, i.e., the two pillars within it and the vaulted system. While the fate of the rear entrances was under discussion, the construction of the pillars above most likely went ahead, which were duotone like those in the crossing and the basilican body. In this regard, there are some particularly interesting documents concerning the construction of a vault between the duotone columns of the apse in February 1260 and the execution of three vaults between two columns in May 1260, after three vaults had already been built. These documents, in the context of a building phase, lead us to believe that the duotone columns of the apse were built first with the relative scaffolding and additional scaffolding and centring to complete the vaults of the nave, and later, after having dismantled these scaffoldings, the six vaults perhaps belonging to the central space of the crypt were built. That same month of May, other vaults were created in the crypt area in front of a side door. The crypt and the apse were completed in 1260, and from 1261, work was carried out on the outside area opposite the three eastern entrances (Giorgi & Moscadelli, 2005, 75–76; Butzek, 2010, 112–114).

During the completion of the apse of the intermediate cathedral, it was decided to reuse the Ionic crypt. The lowering of the crypt floor by about 1.50 m led to some alterations to the

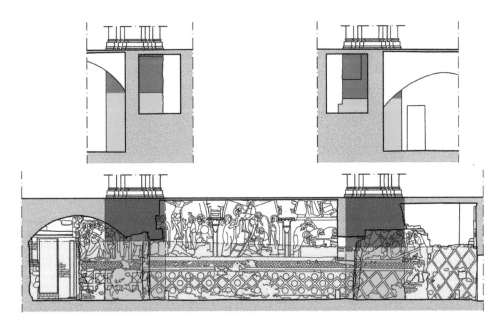

Figure 3.28 Sections of the crypt; relationship between the wall with scenes of the Passion and the foundations of the pillars of the dome; these latter are made up of a part in conglomerate (green) and a part in stone ashlars (red).

pre-existing structures and architectural elements (Figure 3.28). Part of the foundation shaft of the pillars and sections of the foundations of the perimeter walls of this area were made visible; the three original entrances on Valle Piatta were lowered, and it was also necessary to do away with the Attic base of the semi-columns of the counter-façade continuing the shaft up to the new floor level. On the same wall, the two corner capitals have suspended pilasters that end at the bottom with a sort of rounded travertine shelf that corresponds to the floor level of the original crypt: these could be travertine bases reworked into a corbel, in this case, not continuing the shaft of the pilaster down to the ground.

In addition, the foundation wall offsets, as can be clearly seen from the traces now visible in the area close to the corner semi-columns of the southern wall, were also chiselled. Also in this last wall, the arches of the pit foundations, formerly underground, later came to be above ground (Figure 3.29).

When the crypt was rearranged, the overall width of the space was delimited by the three perimeter walls of the original apse, which gave rise to the intermediate cathedral, and by a new brick wall built between the foundation shafts of the east pillars of the dome, on which the scenes of the Passion of Jesus were then painted (Bagnoli, 2003, 107–147; Cervini & De Marchi eds., 2016).

In the reused Ionic crypt, the capitals kept their original height, as can be seen from the levels of the Ionic capitals on the north and south corner semi-columns and the semi-column on the entrance side, and from the level of the north-east corner capital.

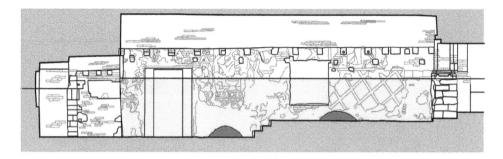

Figure 3.29 Longitudinal section of the crypt indicating the lowering of the floor levels (green) to adapt the earlier crypt to the final structure of the apse of the intermediate cathedral.

Other corbels, as well as a system of small columns placed in the central space between the foundation shafts, were placed to support a probable ceiling of groin vaults and diagonal ribs, of which several traces remain (Giorgi et al., 2009, 5–29).

It was pointed out that the present-day corbels of the north and south walls, with very elongated triangular masonry underneath, would have been too close to the original floor height of the crypt (Haas & von Winterfeld, 2006a, 371–372). In fact, the triangular part visible on the southern side has a chaotic masonry that would suggest it was created in a breach. On the opposite side, at the level of the Ionic capital, which is now missing, a trace of painted plaster remains; in the corner pilaster of the same wall, other fresco fragments can be seen at a higher level than the Ionic capital. All of this infers that the corbels with an elongated triangular base were made during the reuse of the pre-existing Ionic crypt; to confirm this, these corbels are for the most part placed in line with the pillars inside the crypt.

It is clear that the vaulted system of the reused crypt stood on supports that were partly newly built and adapted, such as the foundation sections of the pillars of the choir and dome, and in part added for the occasion, such as the corbels of the west brick wall and the two surviving marble bases of columns placed in the centre of the present-day crypt.

The reworking of the east foundation shafts, in order to create the supports necessary for the vaulted structure, should be ascribed to this same phase; the corner parts to the side of the Crucifixion fresco were demolished, reducing their horizontal section by almost 10% (Figure 3.30). This operation, like the reworking of the octagonal foundation shafts, appears rather reckless when coupled with the fact that the conglomerate foundation shafts had been brought above ground. All of this would seem to confirm the builders' positive assessment of the quality of the mortars, which must have been well known; in fact, not much time passed between the creation of the pit foundations and the decision to reuse the Ionic crypt by lowering the walking surfaces and consequently bringing the foundations of the pillars above ground.

The level of attention in any case paid to this operation is proven by the overlap between the crypt plan and the east pillars of the dome above. The partial demolition of the foundations was in fact executed up to the base of the pillars, therefore taking their presence into account. The adaptation of the foundation shafts, before the vaults were created, was probably carried out when the connection between the pillars and foundations was visible.

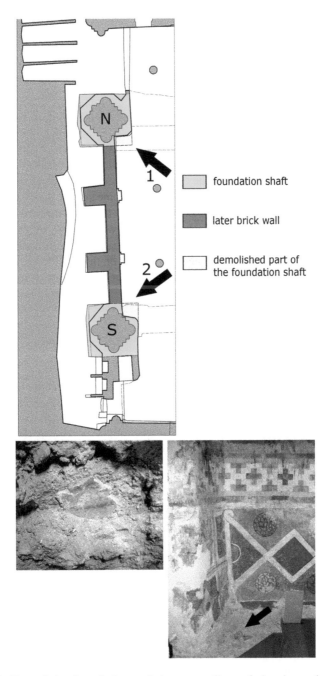

Figure 3.30 (a) Plan of the foundations of the east pillars of the dome (green), between which a brick wall was inserted later on (red); the arrows indicate the demolished parts (yellow) of the original foundations. (b) A fresco fragment can be seen in the cast of a pillar foundation. (c) Detail of the base of the south pillar.

The solution of semi-columns with side pilasters and Ionic capitals is a refined solution that may have been combined with columns rather than the present-day unrefined octagonal pillars which are an adaptation. This seems to confirm the initial intent of a crypt or lower church divided into bays by columns and with the single room of the apse covered by trusses above. The crypt extended to the transept must have been abandoned early as it would have implied the demolition of the apse area of the earlier cathedral. This aspect is linked to the choice to start the construction from the apse and the transept and then move on to the basilican body and conclude with the crossing, by which time the original cathedral could be completely dismantled.

In 1263, several documents attested the purchase of lead for the roof of the dome, as well as the creation of the copper sphere to be placed on top (Giorgi & Moscadelli, 2005, 77). Due to the relationship between the base of the dome and the surrounding loggia, the lead roofing and the sphere had to be placed on an external wooden dome and not directly on the masonry one. Some doubts about the brick dome being completely finished may have arisen due to the subsequent events linked to the "Duomo Nuovo", the New Cathedral, which are discussed in the following pages and to which reference is made. It cannot be entirely ruled out that the masonry part of the dome was not completely finished; on the other hand, the irregularities observed in the profiles of the dome could testify to a discontinuous construction carried out in several episodes.

3.2 The surpassing of the intermediate cathedral

3.2.1 The new façade and the addition of a bay

In 1280–1284, the basilican body was extended to the west with the construction of an additional bay and the present-day façade (Butzek, 2006, 24). Literary sources, in reference to this addition, quote documents on the execution of "simple faces" (Quast, 2007, 106–110), that could imply the sides of the new bay. This expression sometimes is referred to the new façade, and its foundation ceremony is recorded as having taken place in 1284 (Butzek, 2006, 24); in this case, "simple face" should probably be understood as meaning a façade constructed of rustic masonry. The first construction phase of the façade was directed by Giovanni Pisano between 1284 and 1297 and concerned the first level up to the base of the small loggias (Quast, 2007, 110–114; Tigler, 2007). This was followed by the construction of the upper part of the façade between 1297 and 1317, under the direction of the master builder Camaino di Crescentino, with variations regarding the loggia, and the size and position of the central oculus (Haas & von Winterfeld, 2006a, 432–451; Quast, 2007, 114–115).

The construction of the façade must have necessarily preceded the addition of the new bay adjacent to it. In particular, in the north aisle the adaptation in the connection areas established by the façade is evident: the longitudinal arch placed against the north portal is clearly rotated (by 3°) with respect to the alignment established by the pre-existing longitudinal arches.

The deformation of the arch placed up against the previous façade, highlighted by an accurate survey by Walter Haas and Dethard von Winterfel (Haas & von Winterfeld, 1999b, tav. 13), seems to evidence some difficulties in the construction operations necessary to expand the basilican body. The very demolition of the pre-existing façade to create the connection with the new one could have resulted in the deformation of this arch.

The new bay has several unique features. On the sides of the clerestory, a trifora was created with a similar width to the existing ones, however, it was decided not to create at its sides the small arches that made it possible to make the upper wall thicker. Taking the adjacent bay into consideration, the wall of the clerestory was decreased from 1.28 to 1.12 m thick.

The longitudinal dimension of the bay was reduced, maintaining the same height as the vaults in the nave, while in the aisles, the height of the vaults and their longitudinal arches was increased. This resulted in quite a few adaptations, which were sometimes even approximate.

Particular reference is made to the longitudinal arches which, even though they had a shorter span than the pre-existing ones, were raised to a higher height and placed above a sort of dosseret resting on the capitals. It should also be pointed out that the springers of the arch are at clearly different heights; the highest one belongs to the capital of the semi-column standing up against the counter-façade.

Only these arches have a Florentine archivolt, and at their top, they intersect with the base of the cornice made up of double-leaf corbels, alternating with busts of the popes.

The aisles of the new bay were given higher vaults and a much more prominent curvature than the adjacent pre-existing vaults. The vaulted system in the nave was not altered. But the side pilasters, on which the diagonal ribs of the vault rest, were interrupted here by a corbel and a statue placed just below (Figure 3.31).

The greater height of the vaults of the bays adjacent to the side entrances seems to have been required by the extraordinary design of the new façade by Giovanni Pisano compared to the earlier one. It should be said, however, that there are also significant differences between the two new bays: in the south one, the vault and the longitudinal arch are higher. It has been pointed out that the variant nature of the present-day façade is also evident from the position of the buttresses placed between the three entrances, which do not take into account the

Figure 3.31 The longitudinal arches of the aisles are taller than the bay abutting the façade; this causes a partial overlap with the cornice above decorated with the busts of the popes.

Figure 3.32 Garret of the aisle flanking the square; on the internal side of the façade there is a stone cornice that can be attributed to an original two-pitched roof which was at a lower height than the current one.

width of the naves of the basilican body, positioned at the edge of the longitudinal arches (Toesca, 1965, 38). The thrust of these arches is thereby shifted from the central area of the buttresses where we find the largest resistant section.

In the parts of the counter-façade incorporated into the garrets of the aisles, there are over-hanging owl-beak stone bands which delimit the original position of the sloping roofs of the aisles (Causarano, 2017, 100–102). This roof did not have an accessible garret and rested directly on the vaults. On the clerestory side, there was necessarily a water collection system which channelled it to the external side of the aisles (Haas & von Winterfeld, 1999a, fig. 278). Even today, water is still collected in the areas in front of the triforas of the clerestory and chan-nelled to the outside of the sides (Figure 3.32).

3.2.2 The baptistery and the redefinition of the transept and the apse

The subsequent dates provided in the documents, which testify to initiatives concerning the baptistery linked to the complex and structured progress of building the cathedral, should now be correlated.

The intention to demolish the church of San Giovanni was recorded as early as 1296 with the idea of reconstructing it next to the cathedral bell tower. The demolition of the church was attested in the early 14th century (Giorgi & Moscadelli, 2005, 80).

In this period, however, a different decision was made regarding the position of the new church of San Giovanni Battista. It was placed close to the three rear entrances to the crypt and, exploiting the slope of this area down towards Vallepiatta, the apse of the cathedral was extended above the church of San Giovanni (Figure 3.33). The close connection between the two projects was approved by the decisions made by the General Council of the Municipal-ity of Siena in April 1308, which asked the Opera di Santa Maria to guarantee funds for the

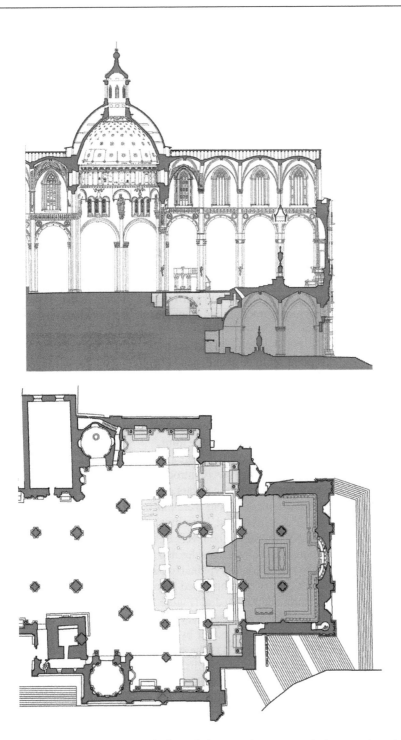

Figure 3.33 Longitudinal section and plan of the crossing area and the apse in relation to the crypt (cyan) and the baptistery (blue).

construction of the baptistery and the enlargement of the cathedral (Giorgi & Moscadelli, 2005, 91).

The documents record some key moments in the construction of the Baptistery: the foundation of the façade facing Vallepiatta in 1317 and the completion of the vaulted system in 1326 which the builders celebrated with a lunch (Giorgi & Moscadelli, 2005, 93). In 1333, the works concerned the extension of the cathedral above the baptistery, and by 1337, they had reached the level of the windows (Giorgi & Moscadelli, 2005, 93–94).

Despite the slope in front of the three entrances to the crypt, large-scale excavations were necessary to construct the baptistery. The longitudinal section shows that the difference in height between the floors of the cathedral and those of the baptistery is about 12.40 m, compared to a height difference of about 5.00 m between the cathedral and the Ionic crypt reused in 1260. This meant that digging of around 7 m was carried out close to the façade with the three entrances to the crypt.

It is not surprising that this intervention damaged the apse, creating cracks that can still be seen today in the sides of the crypt. In order to halt this structural instability, the wall was substantially thickened at the rear façade of the intermediate cathedral, creating a wall in total 3.30 m thick, almost doubling the original wall (Figure 3.34).

The Baptistery and the new choir above had a different alignment to that of the apse of the intermediate cathedral; the pillars were shifted south, resulting in obvious adaptations in the bay adjacent to the crossing.

At the level of the apse, the first bay is in fact skew, particularly on the north side. At the level of the crypt, at this point set to be abandoned, the intermediate octagonal pillars made it necessary to thicken the wall in order to create the support for the apse pillars above. The next two pillars of the apse, for around 1/3 of their plan, were placed outside the enormous wall below, which originated as seen from the thickening of the rear façade of the intermediate cathedral. This position also appears to be an adaptation in that it creates a second bay of the apse which is shorter than the first and the third.

As for the height, in the choir, a nave the same height as the basilican body was introduced, and elsewhere, significant changes were introduced. The aisles were raised by around 1.70 m, while the level of the capitals of the pillars only increased by around 0.4 m. So the heights of the springers of the vaulted system were placed on a sort of dosseret rising from capitals. This solution had already been used, less obviously, in the bay added to the basilican body. The raising of the aisles also raised the levels of the windowsills of the triforas of the clerestory and of the alure that runs along the sides and along the end wall in the nave.

The vaulted system of the reused Ionic crypt had to be dismantled due to the thickening of the octagonal pillars of the crypt, which supported pillars with a different position than the pillars of the apse of the intermediate cathedral (Figure 3.35). In addition, the floor levels of the new apse cut the vaults of the reused Ionic crypt. The drastic resizing of the crypt may have resulted in the repositioning of the pulpit by Nicola Pisano, built between 1265 and 1268 (Figure 3.36) (Tigler, 2009, 122–131). The pulpit, originally placed between the two pillars of the dome next to the bell tower, was disassembled in the third decade of the 16th century and relocated on the two large brick arches that now stand on the foundation shaft of a pillar of the dome and on an octagonal pillar inside the crypt; it could not be repositioned on the filling of the crypt as it was too unstable and subject to settling over time causing potential differential subsidence that could have damaged it (Figure 3.37).

Nevertheless, the builders seem to have reflected more generally on the constructive aspects found at crypt level. A comparable brick diaphragm arch was also created between

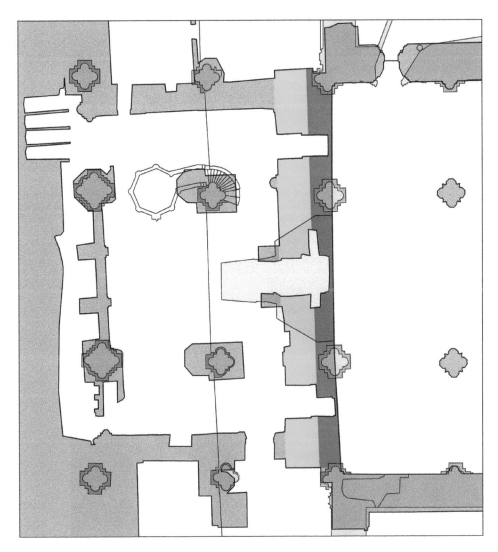

Figure 3.34 Plan of the crypt indicating the rear façade of the intermediate cathedral (green) and the wall thickening (red) added for the construction of the baptistery; the position of the cathedral pillars (cyan) is also indicated.

the remaining foundation shaft of the dome pillar and the other octagonal pillar. Moving on from the earlier situation, in which the pillars that continued in the crypt were probably reinforced by diaphragm arches inserted into the vaulted system of the crypt, to the new situation with the crypt dismantled and filled with rubble, the builders did not want to leave the lower part of the pillars of the dome without constructive protection; so here they decided to protect both bases of the pillars of the dome with diaphragm arches. Near this diaphragm arch, a wall was added to support a barrel vault that covers a residual area of the crypt left accessible (Figure 3.38).

Figure 3.35 Cross section of the crypt with the hypothetical conformation of the vaults of the central part (cyan), reconstructed on the basis of the traces of the vaults that survived above the corbels.

Figure 3.36 The pulpit by Nicola Pisano.

A further extraordinary change was made in the third decade of the 14th century, during the work on the baptistery and the new apse (Giorgi & Moscadelli, 2005, 93), which for a period drew great attention to its design and execution, and therefore most of the financial resources: the construction of the new cathedral, hereinafter referred to as the Duomo Nuovo.

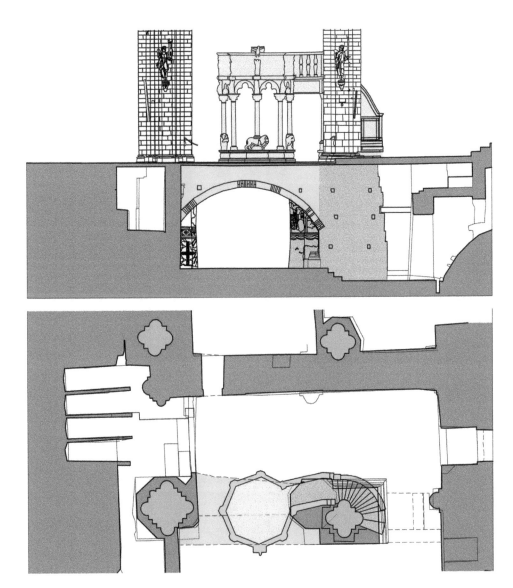

Figure 3.37 Excerpt of the plan and section of the area below Nicola Pisano's pulpit; the section shows the diaphragm arch (cyan) on which the pulpit rests; in the plan, the pulpit (green), diaphragm arch (cyan) and cathedral pillars (orange) are superimposed.

3.3 The Duomo Nuovo: a lost challenge

On 23 August 1339, it was decided that work would start on the Duomo Nuovo, and in September 1339, it was decided that "due carte di pecora" [two sheep parchments] to be used for two "disegnam[en]ti del duomo" [drawings of the cathedral] would be purchased. In the months that followed, other "carte di pecora" were purchased to prepare drawings

Figure 3.38 (a and b) Views of the crypt and (c) detail of the fresco with scenes of the Passion and the arch abutting on it.

(Borgherini, 2001, 61–101). The ceremony of the laying the first stone of the Duomo Nuovo's façade was held in February 1340 (Giorgi & Moscadelli, 2005, 97 note 224). The documents allow us to trace the progress of the works, which started with the façade, now commonly referred to as the "Facciatone" [the large façade], and extended to the naves. The purchase of materials for pillars, masonry, arches, vaults and wooden centring for the vaults was documented; in 1345, work was carried out on the vaults of the nave, for which special bricks referred to as "bottacci" were used (Giorgi & Moscadelli, 2005, 98). Intense work was carried out until the fifties, with the purchase of houses in the area where the Duomo Nuovo was being built and various building activities, including the construction of the main vaults of the nave (1350–1351), which continued unabated until the early months of 1357 (Giorgi & Moscadelli, 2005, 98–101). In these same years (1354/1355), the problem of acquiring an area for the Duomo Nuovo till occupied by part of the bishop's palace was also addressed (Giorgi & Moscadelli, 2019, 178–186).

Now let us examine some practical construction aspects encountered in this new phase of the works, focusing first on an analysis of the designs of two plans for the Duomo Nuovo produced on sheep parchment and on the correlations they established with the earlier cathedral.

In both drawings, the new church is oriented orthogonally to the pre-existing building, assuming the same position as the crossing. The drawings both have a basilican body with three naves, divided into six bays covered by groin vaults (Figure 3.39).

The two solutions clearly differ in the proposal for the apse. One of them had semicircular deambulatory, placed in continuity with the aisles and with twelve chapels facing onto it. On each side, starting from the façade, the first two chapels correspond to bays of the earlier building, while the next chapel, the first of the radial ones, extends beyond the alignment dictated by the existing cathedral. The latter is divided into three sections, of which the external

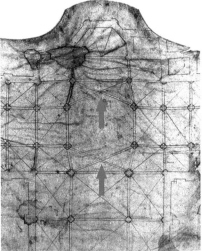

Figure 3.39 Partial plan of the designs for the Duomo Nuovo; the arrows indicate the pillars of the crossing which were set to be demolished and replaced with crossed arches.

ones, the residual areas of the basilican body and the apse, would have become autonomous church spaces. The other solution has a simple semi-octagonal apse, preceded by a three-aisle building with a shallow bay. In this case, the earlier cathedral is fully integrated with the new building and forms its transept.

The two projects share a construction solution for the hexagonal crossing. Both eliminate the north and south pillars of the dome, which were placed along the axis of the nave, and replace them with crossed arches. If this had gone ahead, it would have been very enterprising construction which would have also involved altering the base of the hexagonal drum in order to adapt it to the form of the new arches. The pre-existing arches, with their reduced span, would have been replaced by arches of double the span (about 18 m), much more subject to deformation than the previous ones also because the masonry structure of two crossed arches would have implied discontinuity at the point where the north and south corners of the drum hexagon intersect (Figure 3.39).

It seems unlikely that such a solution could have been imagined with a masonry dome that was entirely built; the dome was most likely only partly constructed, and perhaps temporarily completed with a wooden structure with a lead roof. Unless the plan was to demolish the dome, saving only the east and west pillars of the crossing with a part of the drum. In fact, the dismantling of the dome was a possibility considered in a survey dated 1356 (Giorgi & Moscadelli, 2005, 101), which we shall focus on after having shed new light on the intense activities to construct the Duomo Nuovo, examining the surviving walls south of the present-day transept of the cathedral: the east aisle, the "Facciatone" and a section of the west side, equal to three bays (Figure 3.40).

The aisle is the most intact part with its five bays, three of which are incorporated into the Opera del Duomo museum and two are visible from the outside, which, with their groin vaults, are seen as a loggia in today's urban context (Figure 3.41). Near the connection between this surviving nave and the adjacent southern transept of the present-day cathedral there are significant remains of architectural structures. In the garret of the transept are the remains of a vault of an additional bay of the aisle and part of the arch of a window of the same bay is visible. Moreover, the outer wall of this bay, as evidenced by the end part which extends beyond the roofs, was shortened to create the present-day corner bay between the transept and the choir.

The wall remains just examined show that another rectangular bay, the sixth, had been built at the edge of the trapezoid connecting to the crossing. The sixth bay was partially inserted into the transept of the earlier cathedral. The pillar between the fifth and sixth bays is now incorporated into the southern wall of the transept (Figure 3.8).

On the opposite side, the construction of the sixth bay had only been started. At the base of the east wall of the bell tower, a section of a lobed marble pillar can be seen, and on the same wall, there is a clear repair to a breach in the wall that rises up to the garrets, for a total height of around 16.50 m.

This pillar placed in the wall of the bell tower and the other corresponding one of the opposite bay formed the pair of pillars to be connected by diagonal arches to the south-east and south-west ones of the crossing; the two north-east and north-west branches of the diagonal arches, according to the two plans on sheep parchment, would have had to support the corresponding walls of the hexagonal drum of the crossing, thus eliminating the southern pillar.

The pillar built inside the bell tower wall would have implied longitudinal arches confirming the intention to dismantle the entire bell tower, which in fact is not indicated in the two plans of the Duomo Nuovo.

In the interior elevation of the "Facciatone", the parts corresponding to the nave and the west aisle are highlighted by the marble cladding, while the remaining brick walls were incorporated into the garrets or joined to the vaults. The overhanging fragments of wall that flank the level of the vault of the nave of the "Facciatone" are particularly significant. On the west side in particular, where there is a particularly marked overhang, it is clear that these are the remains of a collapsed wall and not an area where toothing was created for a wall to be built. The walls supporting the vaults of the nave were built on the longitudinal

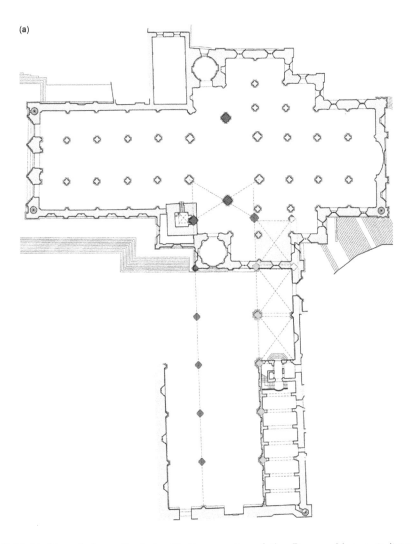

(a)

Figure 3.40 (a) Plan of the cathedral with the remains of the Duomo Nuovo, indicating the perimeter masonry (yellow), the pillars (green) of the surviving aisle, the position of the pillars (red) of the demolished aisle, the collapsed vault (cyan), traces of which remain in the garret, with the pillar that no longer exists (red); part of a pillar (blue) is incorporated into a wall of the bell tower; according to the parchment drawings two of the pillars of the crossing (brown) should have been demolished.

(Continued)

(b)

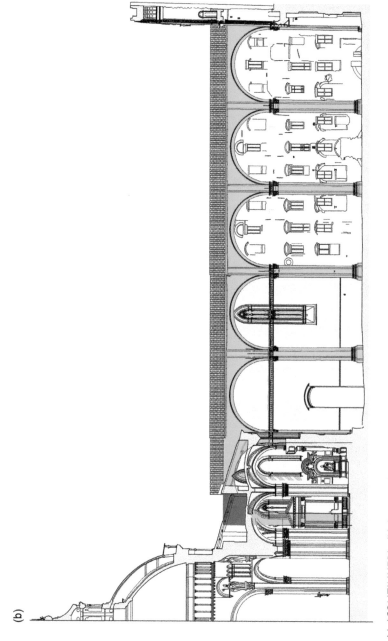

Figure 3.41 (CONTINUED) (b) Longitudinal section of the "Facciatone" of the Duomo Nuovo at the transept of the cathedral; an additional bay of the Duomo Nuovo was positioned within the present-day transept (green), where the remains of the vault and a side window are preserved in the garrets (yellow); a cut in the masonry at the top of the side is also high-lighted (purple).

Figure 3.41 View of the remains of the Duomo Nuovo; the aisle (left), the inner side of the Facciatone (centre), and a section of a side (right).

arches that divided the aisles, as mentioned in the documents, and they might have risen up to the lines of the eaves. Given the necessary construction relationships between the sections, the execution of the vaults of the nave implied that the corresponding vaults of the aisles had already been constructed. Surviving elements of the west aisle include, in addition to the trace in the "Facciatone", the division of the side into three bays built close to the bishop's palace, where there are semi-columns, mullioned windows and the profile of the groin vaults.

The construction of the basilican body, which was almost complete by this point, meant that technical difficulties had to be faced, as well as implications linked to the cost of the works which, in the crossing area, would have involved the connection between the Duomo Nuovo and the earlier cathedral. This aspect was observed with concern by the "Governo dei Dodici" (Government of the Twelve), which had led the Sienese community since 1355. So the masters Domenico di Agostino and Niccolò di Cecco del Mercia were asked for an opinion, dated to 1356 (Giorgi & Moscadelli, 2005, 101; Haas & von Winterfeld, 2006b, 784, doc. 140).

The report makes explicit reference to the demolition of parts of the "old" cathedral: the bell tower, the dome and some vaults. It was thought that it would be extremely expensive to carry out these works in order to integrate the old and the new cathedral, and it would significantly prolong the time required to complete the works, estimated as a century. The two masters proposed that the work on the old cathedral be completed with the extension above the baptistery and that an independent church be created in the new building under construction. At this stage, uncertainties arose in the management of the construction site, in that they reached the point of creating a large breach in one side of the bell tower, where a pillar of the nave was then built.

The size of the basilican body, which had been defined by this point as evidenced by the remains of the naves and the "Facciatone", can be usefully related to the drum and dome of the earlier cathedral. The leap in size between the two buildings is extraordinary. The height of the longitudinal arches of the nave was equivalent to that of the arch of the crossing near the apse of the earlier cathedral. But with the height of the vaults of the nave, it even reached the top of the dome (Figure 3.42).

In light of this, in the initial design phase of the Duomo Nuovo, with the two alternative planimetric solutions under discussion, it was still possible to consider saving the drum and dome of the earlier cathedral. In fact, this intention could be seen in the two drawings mentioned above in the massive arches drawn with a double line close to the crossing, which cross at the north and south pillars that were to be dismantled, while in the remaining bays, the ribs of the groin vaults are drawn with a simple line. However, given the imposing mass of the built basilican body, the dismantling of the drum and the dome seems inevitable, which not surprisingly had been envisaged in Domenico di Agostino and Niccolò di Cecco del Mercia's report in peremptory terms, leaving no room for compromise solutions (Tragbar, 2018, 45–48).

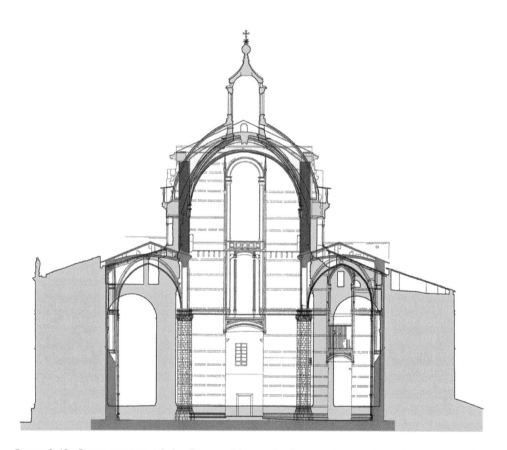

Figure 3.42 Cross section of the Duomo Nuovo (red) superimposed on the cross section of the dome; the nave of the Duomo Nuovo rises up to the base of the lantern of the dome.

In the spring of 1357, the condition of the structures of the naves of the Duomo Nuovo became critical to the point of requiring expert opinions, including those of the Florentines Francesco Talenti and Benci di Cione (Giorgi & Moscadelli, 2005, 102 note 247). An interesting written opinion by the latter has been conserved, which focuses on the numerous critical points concerning the entire layout of the Duomo Nuovo (Haas & von Winterfeld, 2006b, 784–785, doc. 141).

The opinion focuses on the serious static conditions of four columns, and suggests they be demolished, together with the arches, vaults and masonry resting on them: the vaulted structures are described as broken and the walls folded. So overall, it suggested dismantling nine bays.

Benci di Cione prepared a diagnosis of the structural damage. He found defects in the construction of the pillars, made up of a rather thin marble facing and too much bad quality filling. He found the connections between structures to be deficient and the asymmetry in the composition of some masonry to be damaging.

He proposed the full reconstruction of the pillars using marble ashlars and suggested the use of transversal ties in the aisles and nave. He recommended constructing walls above the arches that were no wider than the arches themselves; the external squared buttresses at the level of the clerestory had to be interlocked into the walls and have dimensions consistent with the pillars below.

Benci di Cione thought it would be ineffective to repair or thicken the pillars; he suggested reinforcing the arches by adding arches at the intrados. However, he concluded that the vulnerable state of the structures was so serious that it would be preferable, even considering the cost, to reconstruct the four pillars and therefore the nine bays converging on them.

In May 1357, another panel of experts was called upon to assess the stability conditions of the structures. In this case too, it was thought that any reinforcement of the structures would be ineffective, and it was decided to dismantle each part of the Duomo Nuovo, except for the perimeter walls (Haas & von Winterfeld, 2006b, 785, doc. 142).

The critical conditions reached by the structures are still seen today in the surviving aisle, examining the profiles of the pillars and the section of semi-columns found between the longitudinal arches. From the base to the capitals, there is a tilt towards the nave that varies from 13 to 28 cm; starting at the capital, the semi-columns tilt in the opposite direction which in the short wall remains vary from 12 to 25 cm (Figure 3.43). The remains of the aisle therefore evidence the deformation of the structures with two different tilts. The extent of the deformation of the short section of the semi-column, of around 6 m, is striking if we consider that the masonry continued upwards for a further 13 m, with an increasingly marked overhang towards the outside. With respect to a similar situation, it is easier to understand why Benci di Cione in his report referred more than once to the fact that the walls were warped.

The collapse may have occurred within a few days that same May, as in this period the accounting records of the Opera show that the work was unexpectedly interrupted. This was confirmed the following July, when the labourer Ceccarello di Guido was instructed to dismantle arches and walls, and to free the area from debris (Giorgi & Moscadelli, 2005, 102).

It has been pointed out how, contrary to the claims made by a part of historiography, the current state of incompleteness of the Duomo Nuovo cannot be attributed to the consequences of the plague in 1348, as intense work continued in the following years (Giorgi & Moscadelli, 2005, 98 note 233; Butzek, 2006, 63 note 782). The structures surviving today are actually the remains of a basilican body that was almost completely built, but then fell into its current conditions after collapses and the dismantling of the unsafe parts.

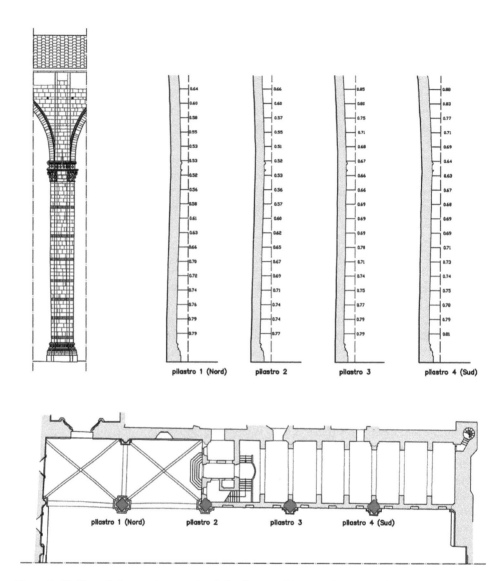

pilastro 1 (Nord)   pilastro 2   pilastro 3   pilastro 4 (Sud)

Figure 3.43 Plan of the surviving aisle of the Duomo Nuovo, elevation of a pillar and profiles of the pillars showing the extent of the tilts.

The great constructive gamble of the builders was to take some key aspects of the basilican body of the earlier cathedral as a construction model. Similarly to the latter, they decided against the use of ties and effective buttressing systems, which were limited to modest pillars placed on the outside of the sides and the clerestory. The comparison between the cross sections of the two buildings, made possible by the configuration of the aisles testified by the interior side of the "Facciatone", is particularly indicative; it went from a nave roughly 9.00 m wide and 24.95 m high (considering the intrados of the vaults joined to the transverse arches), to a nave 14.70 m wide and 34.20 m high. A further non-secondary aspect was that in the nave, the ratio between the width and depth of the bays was reduced; this choice, the span being equal, tended to increase the surface of the bays and therefore to accentuate the magnitude of the thrust transmitted by the vaults of the nave. This extraordinary dimensional jump made in the Duomo Nuovo, without however incorporating constructive protections to contain the thrusts that such a choice made inevitable, turned out to be fatal in the end (Figure 3.44).

The works were immediately resumed with the completion of the choir under construction in the area above the baptistery. The pillars, vaults and windows of the nave were created. The works necessary to create the new apse to the crossing and to create the new transept were started between 1359 and 1360. This involved the demolition of the walls of the apse and a large part of the transept of the intermediate cathedral (Giorgi & Moscadelli, 2005, 103–104), of which some fragments were most likely saved for use in the chapels on the north side (Haas & von Winterfeld, 1999b, tav. 3).

The connection between the crossing and the addition was not without its problems. The vaults between the pillars of the crossing and the perimeter walls of the apse and transept area were dismantled in some cases and in others they collapsed during the works. A large purchase of bricks was recorded in 1362 and the registration of workers' payments for the construction of the vaults continued until 1366 (Giorgi & Moscadelli, 2003, 104 note 258; Giorgi & Moscadelli, 2005, 104–105), when the entire vaulted system of the apse and transept must have been completed.

Between 1358 and 1360, some works were carried out on the dome (Giorgi & Moscadelli, 2005, 104, 105 note 264); however, it is not entirely clear if this work was required after damage caused by the demolitions and collapses of the structures adjacent to the crossing, belonging to the intermediate cathedral, or if they were works to complete the dome. Any damage to the dome must have been linked to the spread of instability in the pillars and arches of the crossing, which were positioned close to the demolished and collapsed parts. However,

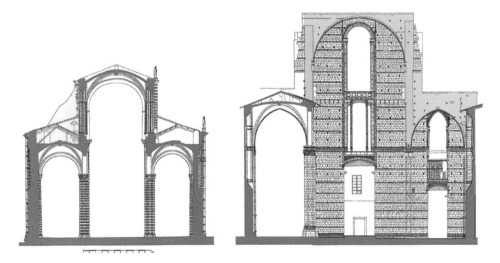

Figure 3.44 Comparison between the cross sections of the aisles of the cathedral and of the Duomo Nuovo highlighting (green) the edge of the vaults of the nave and of the aisle.

the pillars and arches of the crossing, insofar as intact and free of significant deformations, would suggest it was work to complete the dome.

The completion of the work phase also coincided with the final backfilling of the reused Ionic crypt, except for the south aisle, which was covered with a brick barrel vault and made accessible by a door created by breaching the wall on the south side (Giorgi & Moscadelli, 2005, 104) (Figure 3.45). It cannot be ruled out that other parts of the crypt were accessible up to the 17th century (Leoncini, 2005, 459–482).

Figure 3.45 The underground rooms of the cathedral, known as the "crypt of the statues", before the excavation work that uncovered the current crypt.

The non-standardized sizes of the perimeter walls of the cathedral would seem to reflect the orography of the place rather than the complexity of the construction history. For instance, there is a 12.25 m difference in height between the entrances to the cathedral and the baptistery. The sides of the basilican body and the south end of the transept, all at the level of the square, vary in thickness from 1.57 to 1.80 m. The remaining walls of the transept and those of the apse, with foundations at lower levels, were built with thicknesses varying from 2.00 to 2.35 m.

The works to the façade of the baptistery nevertheless greatly reduced the wall thickness, which should be considered. The part clad with thick marble is interrupted just above the gables of the windows above the portals and, at the level of the clerestory of the apse, it continues with plastered masonry which is half as thick (Haas & von Winterfeld, 1999b, tav. 12).

It is evident that this choice reflected the difficulties of the construction site after the events of the Duomo Nuovo; however, from a vertical structural discontinuity perspective, the situation created in the walls is unique and may have resulted in a condition of local vulnerability.

The architects seem to have considered the thrusts from the large arches of the crossing, where the drum and dome are placed, to be safe as they act on the very thick part of the baptistery façade.

The plan at the clerestory level shows that the façade was built with a similar wall thickness to that of the clerestory walls, with which it is placed in continuity (Haas & von Winterfeld, 1999b, tavv. 9, 15). In the basilican body, in the clerestory there are similar wall thicknesses which, however, abut the much more massive west façade. This difference may have influenced the behaviour of structures during earthquakes, causing different levels of damage, as we will see later.

The characteristics of the vaulted system between the crossing, transept and aisles of the apse is unique in that it is made up of bays of a similar height that do not respect the usual basilican scheme, with a nave rising above the side ones. The hall church space thereby created, with springers on dosserets rising from capitals, anticipates solutions created in the second half of the 15th century in the cathedrals of Pienza and Perugia and in San Fortunato in Todi (Rocchi Coopmans De Yoldi, 2006c, 154).

This created clear construction asymmetry between the longitudinal and transversal axes. In the longitudinal direction, the sides of the nave form a sort of large diaphragm arch, partially interrupted by the trifora of the clerestory, which rises up to the loggia at the base of the dome. In the north-south orthogonal direction, the counteracting structure is the vaulted system of the transept at a considerably lower height, placed just above the arches of the pillars of the crossing (Figure 3.46).

Tests were carried out on these vaults to check if any special devices were used, such as diaphragm arches, which would improve the effectiveness of the counteracting action. Checks were conducted in both arms of the transept, in correspondence to the perimeter arches of the bays.

In the south transept, the test (S. 1) carried out on the transverse arch abutting the south pillar of the dome highlighted a diaphragm arch, made up of an entirely brick wall around 45 cm thick. The bricks at the top are arranged in three parallel rows longitudinal to the arch below. A further test was carried out above the pillar of the triangular bay that abuts the south-east side of the drum (S. 2) in order to check the meeting point of two diaphragm arches positioned in a north-south direction and abutting the eastern arch of the crossing. Here too the masonry is made of bricks, with bricks 7 cm thick at the top and arranged transversally to the arches below.

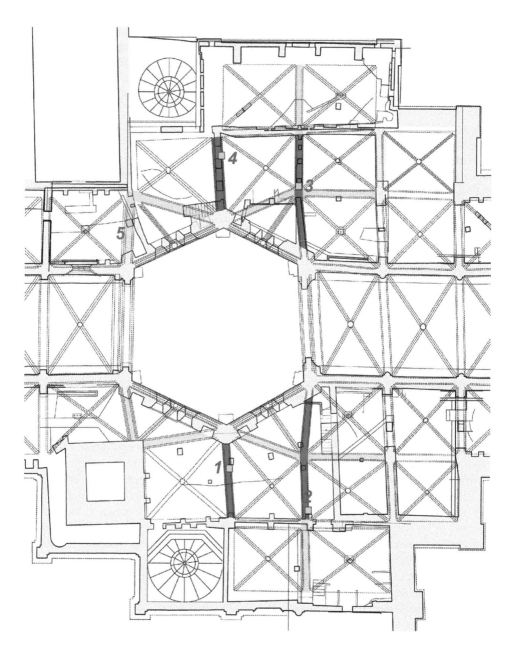

Figure 3.46 Plan of the garret surrounding the crossing indicating the system of diaphragm arches between the vaults, partly verified (red) with tests (1–4), partly hypothesized (cyan).

The use of diaphragm arches, featuring some variations in the masonry technique, was confirmed in the north arm of the transept. In the area above the pillar of the triangular vault on the north-east side of the drum (S. 3), a diaphragm arch was identified, extending

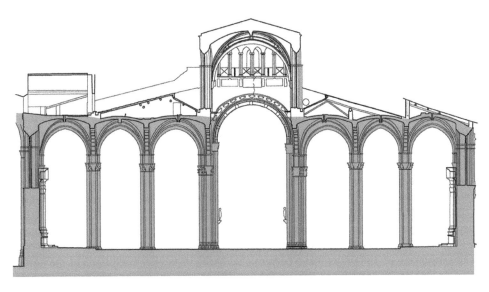

Figure 3.47 Cross section of the transept area with seven bays highlighting (blue) the diaphragm arches that flank the arch of the crossing.

to two arches which, in the north-south direction, abut the eastern arch of the crossing. The
masonry, 55 cm thick overall, had two half-brick facings and was made up of rubble with
an inner core of good consistency conglomerate made up of lime mortar, large particles
of rounded aggregates and brick fragments. The laboratory analysis of the filling material
revealed the heterogeneous characteristics of the aggregates used, abundant fragments of
cocciopesto, carbonate rocks, sandstone and serpentine; lime mortar was also used (Garzonio et al., 2010, 117–118). In the transversal arch abutting the north pillar of the dome, the
test (S. 4) revealed a brick diaphragm arch approximately 60 cm thick.

The top part of all the diaphragm arches discovered is just below the screed at the extrados
of the vaults; so the masonry in correspondence to the keystone of the arches is approximately
0.9/1.10 m high. The diaphragm arches constructed at the edge of the bays, covered by groin
vaults and triangle vaults, create a framework that considerably strengthens the entire vaulted
system, making it possible to effectively counteract the thrusts of the arches of the crossing in
the transept area. It must be taken into account that in the eastern part of the transept, between
the pillars of the dome and the perimeter walls, there are up to three bays. In this case, in
particular, the diaphragm arches perform a counteracting function and a connection between
the pillars and the perimeter walls, which assume fundamental importance (Figure 3.47).

3.4 Earthquakes and the re-evaluation of the construction concept of the cathedral

The numerous, and at times violent, earthquakes resulted in the use in Siena cathedral of all
the construction devices that medieval builders had chosen not to use: buttresses, chains and
flying buttresses. It was a gradual and discontinuous process and not the result of a preventive
analysis of the construction aspects of the dome and its vulnerabilities. Rather, decisions

were based on conditions of necessity with respect to the serious damage suffered by different parts of the building in several episodes between the 17th and 18th centuries. These unrelated interventions as a whole, also carried out at considerable time intervals, became widespread and affected the whole building. The aim pursued appears to be conceptually unitary and seeks to strengthen the structures of the cathedral; however, clear compromises and contradictions were also made in the search to mitigate the effects of these interventions on the architectural structures of a building of such iconic status as Siena cathedral.

An earthquake that damaged the roofs and vaults of the aisles occurred in 1558. The filling of cracks that opened up in the vaults was recorded at the end of April that year (Sembranti, 1997, 70 note 13). The initiatives carried out after the earthquake saw the involvement of Cosimo I de' Medici, who ordered the architect Bartolomeo Ammannati to assess the work that would be required. Ammannati carried out inspections to check the damage, but nonetheless we do not know the extent of it. The main measure he suggested was to introduce ties, but there are no reports that this work was done (Sembranti, 1997, 68–74). During the in-depth analyses of the building we did not identify ties that could be attributed to Ammannati's intervention, which suggests that the Florentine architect's plans were not carried out.

The earthquake of 1691 must have been related to the instability conditions of the north-west aisle, which in 1694 were considered so critical that urgent intervention work was required. This work was carried out in just a few months that same year (Sembranti, 1997, 77–78).

The examination of the off-plumb elements of the buttresses on the north-west wall of the clerestory highlighted the vertical alignment of the buttress adjoining the drum, as could be expected; then however, moving towards the façade, the tilt to the outside of the four intermediate buttresses gradually becomes more pronounced, following the sequence 2, 4, 6 and 5 cm. The greatest inclination, starting from the façade, is therefore at the second bay (Figure 3.48). The cross section of this bay shows that the tilt to the outside continues in the wall below the buttress up to the capital of the pillar; this latter is instead for the most part vertical (Haas & von Winterfeld, 1999b, tav. 22).

Supplies of bricks that would be used to construct the buttresses were documented in 1694, and the reports also mention the cracks that had developed in the vaults of the aisle (Sembranti, 1997, 78).

The decision was made to reinforce the walls of the clerestory with substantial walls 75/80 cm thick, which rise from the garret at an inclined profile, straight for the first stretch and slightly curved higher up, almost up to the top of the original marble buttresses.

The brick buttresses are positioned on the diaphragm arches placed between the groin vaults below; they have a base as wide as the garret, equivalent to the aisle below, and they were interlocked with the pre-existing marble buttresses creating several breaches in the latter.

It certainly can't have been easy to accept the solution of large brick buttresses emerging from the roof, which due to their size stood out against the duotone stone wall of the clerestory.

It is significant that even after the seismic resurgence in the following thirty years, with the earthquakes of 1697–1698, 1701, 1705, 1724 and 1726, with an estimated intensity on the Mercalli scale of VII (1726), VIII (1697–1698) and IX (1724) (Sembranti, 1997, 79 note 51; Consiglio nazionale delle ricerche, 1985, 43–44; Benucci, 1986, 196), similar structures were not created on the opposite side of the basilican body, thereby abandoning the idea of a symmetric counterthrust system. The south-east wall of the clerestory leans more towards the outside than the one on the opposite side. In the second bay, starting from the façade, the

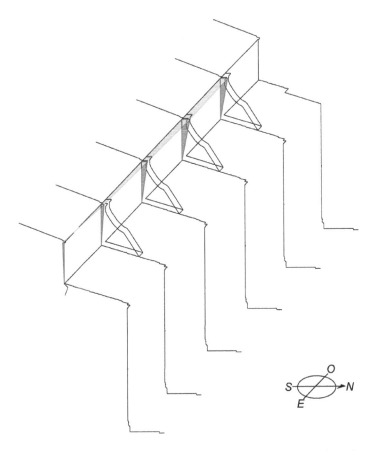

Figure 3.48 The tilt of the west wall of the clerestory is more marked in the area of the second bay.

side of the nave and the clerestory wall now have an inclination of around 10 cm, while in the fourth bay, adjacent to the crossing, the structures are essentially vertical.

The fact that the buttresses would have been highly visible on the cathedral square side was most likely the reason they were not used. With the works endorsed by Pope Alexander VII Chigi, the old episcopal palace was demolished in 1659, which had extended up to the south side of the cathedral. This created the present-day layout of the square (Butzek ed., 1996), with the façade visible along with the view of the side of the basilican body with its duotone stone and the triforas positioned at regular intervals along the clerestory.

The problem of the thrust action on the cathedral square side was tackled by an intervention that found a middle ground between the structural requirements and the visual impact on the building. The roof of the aisle on the cathedral square side has a sequence of closely set wooden struts between the clerestory and outer wall. On the external side, four struts were placed in the space between the windows, which is 4.5/5.0 m wide, resting on pillars that stand on the wall offset of the garret. The strut close to the bell tower is the only one without a supporting pillar on the outside, but here, the wall abutting the base of the bell tower was used (Figure 3.49).

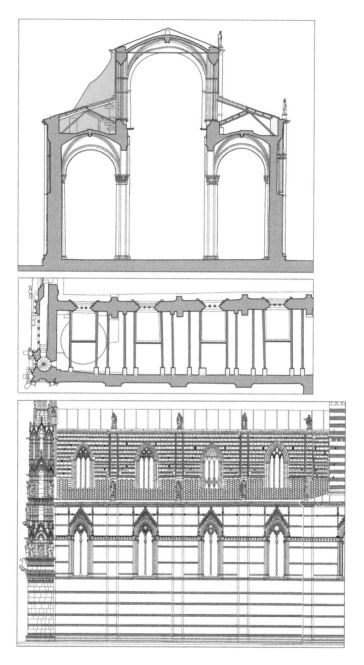

Figure 3.49 Cross section of the aisles, plan of the garret of the right aisle and elevation of
the clerestory on the same side; the drawings highlight (a) the buttress (pink),
(a and c) the tie that skirts the transverse arch and anchors to the long anchor bars
(red) flanking the external buttresses, (a and b) the shoring system of the right
side of the clerestory based on wooden beams (brown) resting on pillars (yellow)
joined to the outer wall.

During the reinforcement works carried out on both sides of the nave, it was necessary to alter the roofs of the aisles, choosing a single-slope roof and leaving the double-pitched roof only in correspondence to the triforas (Giorgi & Matracchi, 2010, 142, fig. 21).

Between 1725 and 1726, work was carried out on the basilican body to shore up the structures damaged by the earthquake of 1724, and interventions using "iron spanners" were set up in the vaults (Sembranti, 1997, 79–80).

Those in charge of the Opera del Duomo considered using visible ties in order to prevent the damage from becoming worse, which was seen in particular in the second transverse arch of the nave. After having consulted several architects, the survey was assigned to Pietro Hustini from Rome, who was already working in Siena on the completion of the Chigi Zondadari palace. Hustini prepared a report and drawings confirming the need to insert visible ties above the springer of the transverse arches. The Opera del Duomo approved the project on 17 June 1726 stressing that the other experts consulted had confirmed the correctness of the solution proposed (Sembranti, 1997, 82–83).

However, regardless of what the documents say, doubts on the choice of inserting visible ties remained. The blacksmith Giuseppe Manganelli was only paid for the work to insert ties, bars and brackets in 1730, and the same was true for the decorator Andrea Andreini who painted the four arches after the ties had been installed. But in the end, it was decided to position them just above the keystone of the transverse arches in order to conceal them. The completion of this work was celebrated in a text painted at the edge of an archivolt of a transverse arch, which sums up the salient aspects: "IN QUESTI QUATRO ARCHI VI E STATO MESSO [DU]E CATENE PER ARCHO PER LA ROVINA CHE MINACIAVANO AL TEMPO DEL ILUST. SIG.RE CLAUDIO BARBAGLI RETTORE NEL 1730" (In these four arches [tw]o ties were placed per arch due to the threat of collapse at the time of the honourable Mr. Claudio Barbagli, Rector, in 1730) (Sembranti, 1997, 82–87).

Each of the four transverse arches of the nave is supported by two ties at the cornice crowning the clerestory. However, even though the ties were positioned at the highest point allowed by the sides of the aisles this did not prevent them ending up below the keystone of the arches. So in the central part of the arches it was necessary to hide the ties with horizontal cladding skilfully concealed by correcting the profile of the intrados and reinforcing the effect of continuity in the profile of the arch by using bands of pictorial motifs, volutes and pateras in the archivolts and coffers on the intrados with a metal flower applied in the centre (Figure 3.50).

The ties were placed in a similar position, but this did not help to create a structurally effective protection, so special anchors were required. Each tie is connected to an external visible iron post placed next to the buttress, which continues down to the level of the springers of the transversal arches. An additional connection was created in this latter position, probably consisting of a tie attached to an anchor bar placed just below the surface of the inner side of the wall.

The date 1728 was painted just to the side of the springer of the transversal arch closest to the drum. The two ties purchased in 1726 were likely installed in this year, after it was decided that they would not be visible, and in 1730, it was likely decided to extend this same construction solution to the three remaining arches of the nave.

The scaffolding set up for the restoration work made it possible to observe differences in the characteristics and dimensions of the anchor bars on the north side. Starting from the façade, the first two arches have anchor bars of the same section (4.7 × 4.7 cm). The third arch has an anchor bar with a section of 6.7 × 4.7 cm. The anchorage of the fourth arch is the

Figure 3.50 Side of the cathedral adjacent to the square; the buttresses of the clerestory
are flanked by the anchor bars of the ties that skirt the keystone of the transverse arch
of the nave.

Figure 3.51 Detail of the keystone area of a transverse arch of the nave (a); in order to conceal the ties flanking the arch, the central part of the intrados of the arch was clad; (b–d) details of brackets and metal connections supporting the cladding.

most unique: one anchor bar is not visible while the other has two paired metal elements, respectively, of 9.5 × 2 cm and 3.8 × 3.8 cm.

The breaches for the lower anchors are highly visible in the stone facing and, where the masonry is visible, they were always repaired with bricks. Where the anchor areas came up against a putlog hole, this passageway was used and the ties placed at the edge. This meant they could avoid creating a large breach, but it resulted in clearly inclined posts.

Another unique aspect of the ties was the cladding necessary to conceal their position below the keystones of the arches (Figure 3.51). One section of the central area has horizontal cladding that gradually connects to the intrados of the arch. The central area is flanked by two brackets and an additional bracket was added on each side lower down; all the brackets are anchored to the tie above using a hook. The area of the arches with cladding is very large and most likely extends for a stretch even beyond the outermost brackets. The concealment of the chains, flanked by arches, also led to the cladding of the area above the ties themselves, while the tooth created below them was attenuated with a mortar connection.

There is the question of which material was used to make the cladding, especially for the intrados area. The lack of obvious cracks would suggest that a lath and plaster frame was made, which was then anchored to the metal brackets and plaster applied on top. In any case, it would be worth identifying the material used; if it was masonry cladding an earthquake could have caused it to partially detach, with consequent risks not so much concerning the stability of the structures but people's safety.

After slight damage to the roofing caused by the 1737 earthquake (Sembranti, 1997, 87 note 76), it was the earthquake of 26 May 1798 that caused considerable damage in the apse area, which not only required extensive reconstruction work but also led to a profound change in the construction concept of the cathedral in the area of intervention. All of this resulted in very lively debate on the effectiveness of the proposed solutions and their visual impact; consequently, the solution implemented stemmed from disagreements and the work of several builders (Sembranti, 1997, 87–108).

Ambrogio Soldani, professor of mathematics at Siena University, drafted a report on the damage caused by the earthquake, which was published in 1798. He described the damage suffered by the cathedral as follows:

> The cathedral was notably damaged by the horrible tremor, especially in the vault of the Choir; and although the great dome and the middle nave did not suffer much, the other two naves were greatly damaged, and the vaults were extremely cracked. It should not be omitted here that at least externally the eastern part [the apse, editor's note] was battered much more than the opposite side.
>
> (Galli, 2006, 349–350)

Although the comment is succinct, it shows that the protections set up in the basilican body were effective; the apse was the most affected area, partly due to the absence of reinforcement structures as well as the fact that it is above the baptistery. In this part, the overall height of the structures is considerably higher than the basilican body. Another aspect not to be overlooked should also be the significant reduction in the thickness of the wall of the top part of the baptistery façade, corresponding to the level of the clerestory.

A detailed report on the damage and restoration of the cathedral was published in 1802 by Vincenzo Meini, Head of Budget at the Opera del Duomo (Galli, 2006, 364–395). Giuseppe Del Rosso and Bernardino Della Porta were given an initial assignment to set up shoring structures that would ward off the collapse of the nave of the apse. Beams were placed transversally above the capitals of the pillars in order to stop the movement caused by the thrusts of the vaulted structures of the aisles. Other shoring structures affected three transversal arches of the nave. The inspection of each part of the vaults of the latter also required a wooden structure (machine) to be designed with stairs and levels, and wheels that allowed it to be moved (Sembranti, 1997, 88–89).

Once the structures had been made safe, the governor of Siena Vincenzo Martini commissioned the architects Giuseppe Salvetti, Giuseppe Del Rosso and Bernardino Della Porta to conduct a survey. On 3 September 1798, they presented a report in which they examined the damage, made a diagnosis and proposed interventions (Sembranti, 1997, 90–94). Del Rosso and Della Porta's involvement in the post-earthquake events in Siena was very extensive and dated back to the previous month of May, when at the request of Governor Martini and by order of the Grand Duke of Tuscany they were tasked with preparing a report indicating the measures necessary to restore the buildings damaged by the earthquake in Siena (Gennari, 2005, 162–166). Salvetti, as an expert of the Grand Duke, had also been involved in the reconstruction of the city since July (Gennari, 2005, 168–172).

In the basilican body of the cathedral, the technicians in charge had not detected any critical conditions, although some cracks had opened up, which in some cases had already existed and had been repaired. In this part, the works carried out earlier had clearly improved the efficiency of the structures.

The damage in the apse was considerable. The pillars of the nave, just above the springer of the arches of the aisles, had bent over and split into two sections. The lower part was out of plomb on the inside, while the remaining upper part tilted outwards. The inclination of the two sections was about 9 and 11 cm, respectively. As a result, the transversal arches were greatly deformed and had lowered and split into pieces, which sometimes had very small contact areas.

They attributed the inward displacement to the lack of ties in the aisles, the reduced size of the pillars and the loads, defined as "extraneous weight", due to the roofing. The outward tilt of the top parts of the pillars, caused by the thrust of the vaults of the nave, was related again to the lack of ties, but in this case, the absence of a buttress was also an issue.

They suggested attaching a tie to the springer of each of the transversal arches of the aisles, recommending that it be anchored to the pillars by a band, avoiding drilling that would have further weakened the pillars.

They also planned for ties to be attached to the same arches at the level of the keystones. In this case, the ties would be anchored to two vertical posts next to the semi-columns on the side of the nave; to strengthen the semi-column, the anchoring system included an additional post in the middle of it. These posts, in turn, were blocked at the springer of the arches with a sort of hooping partly incorporated into the masonry. Between the posts the plan was to create two additional horizontal links, one at an intermediate height and the other approximately at the level of the paired ties. The three vertical posts were to be placed flush just beneath the surface of the masonry (Sembranti, 1997, 130–131, figs. 9–10).

It was considered necessary to create external buttresses for the nave "for the entire height of the church," but the recommendation was to create them with a duotone facing that would connect with that of the adjacent walls. Each of the transversal arches of the nave should have been joined by two ties placed at the height of the keystones (Sembranti, 1997, 93 note 99). In this regard, explicit reference was made to the system of reinforcements in the nave of the basilican body, which is also found in the solution of ties supporting the arches of the aisles, with double anchoring at the semi-columns. The restoration of the vaulted system of the nave was subject to a careful assessment of the degree of damage, to be confirmed as soon as it was possible to erect scaffolding. The use of "wall to wall" beams was recommended for the roofs, avoiding the use of short beams that would have required pillars resting on the vaults.

In April 1799, due to the French occupation which lasted from March to June of the same year (Galli, 2006, 354), the Sienese architect Bernardino Fantastici was appointed construction site manager, and he assumed a critical stance with regard to the proposals of his predecessors. The major disagreement concerned the idea to put visible ties in the aisles, citing "aesthetic damage" as the grounds against them. As regards the previous proposal, he opted for the ties supporting the transversal arches at the level of the keystones and the same anchors at the semi-column of the nave, planned by Salvetti, Del Rosso and Della Porta (Figures 3.52 and 3.53). Fantastici is also thought to have reconstructed the six semi-columns that were seriously damaged by the bending of the wall in the section between the capitals of the pillars and the cornice at the base of the clerestory (Sembranti, 1997, 95–96).

In the nave, the works involved the reconstruction of the vaulted system, which was provided with a tie at the height of the keystones of each of the transversal arches. The affixing of ties at the same time the vaults were constructed meant that the tie could be integrated with the inclined sections of bars that cross the haunch areas of the vaulted system. The horizontal and inclined sections of the ties were connected by anchor bars to the same post, also joined to the stone buttress.

In the reconstruction of the vaulted system of the nave, Fantastici created a distinct masonry structure. In the transversal arches and diagonal ribs, he inserted stone ashlars interlocked with the brick webs (Sembranti, 1997, 97).

While works were being carried out on the vaulted system, after Siena was once again under the governance of Florence, following an inspection, Del Rosso complained about choices that were incompatible with what he had decided with Salvetti and Della Porta. The lively debate that ensued from this conflict led the Secretary of State and Finance to ask the Florentine engineer Neri Zocchi for his opinion (Figures 3.52 and 3.53). The latter proposed sticking with the original project, flanking the transversal arches of the nave with not one (as had been done) but two ties and inserting a visible tie in the central aisles. After various delays, it was decided to task Zocchi and Salvetti with producing a new survey, which was drawn up on 1 March 1800. The works carried out by Fantastici even in the period when they were being contested prevented him from adding a second tie in the arches of the nave. This limitation led Zocchi and Salvetti to arrange for the construction of robust flying buttresses in correspondence to the transverse arches (Figures 3.54 and 3.55).

The transverse arch of the apse closest to the crossing, flanked on each side by two bays, had a flying buttress with an intermediate support on a pillar and connected to the perimeter wall by an additional buttress incorporated into the garret. The remaining flying buttresses, flanked by just one bay, rest directly on the external walls. The sequence of festoon flying buttresses emerging from the roofs is interrupted at the façade of the baptistery, interposed with less emphatic triangular buttresses (Figure 3.56).

The decision to use flying buttresses partially incorporated into the garrets, which directly connect the sides of the nave to the perimeter walls and, in one case, to pillars where an additional arcade creates a connection with the side wall, should be pointed out. In fact, the solution adopted in the basilican body, with the buttress resting directly on the diaphragm arches of the aisles, could not be achieved in this case. The counter-action structures in the apse follow the width, of around 1 m, of the original stone buttresses at clerestory level; but the underlying transversal arches of the aisles are considerably less wide, around 0.70 m. Directly resting on the diaphragm arches, the counter-action structures would have also partly weighed upon the vaulted system, hence the choice to use flying buttresses.

In the apse, unlike in the basilican body, there was no hesitation in creating visible counteracting wall structures on both sides, relying on the fact that they would be hardly visible from the square in that they were preceded by the drum and dome, the bell tower and the surviving aisle of the Duomo Nuovo. Moving on to the area in front of the baptistery façade, the level of the square makes it impossible to see the structural protections emerging from the roofs of the apse.

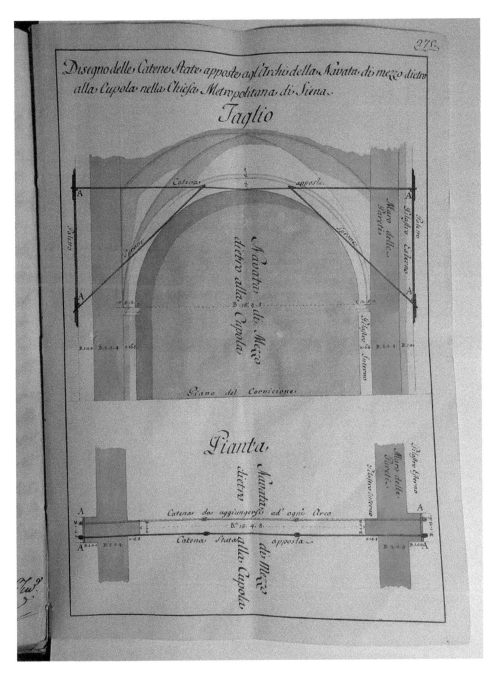

Figure 3.52 AOMS 1595 (932) 275r, "Disegno delle catene apposte agl'Archi della Navata di mezzo dietro alla Cupola della Chiesa Metropolitana di Siena" [Drawing of the ties affixed to the Arches of the middle Nave behind the Dome in the Metropolitan Church of Siena]; drawing by Neri Zocchi (8 November 1799); ties positioned in the nave of the apse with a proposal for their integration.

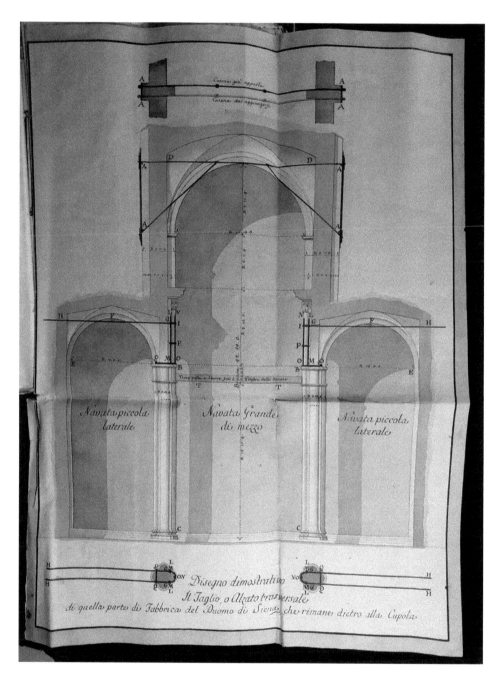

Figure 3.53 AOMS 1595 (932) 295bis, "Il Taglio, o Alzato trasversale di quella parte di Fabbrica del Duomo di Siena, che rimane dietro alla Cupola" [The Cut, or transversal Elevation, of that part of the Building of Siena that remains behind the dome]; drawing by Neri Zocchi (9 December 1799); design of the ties to be placed in the aisles of the apse.

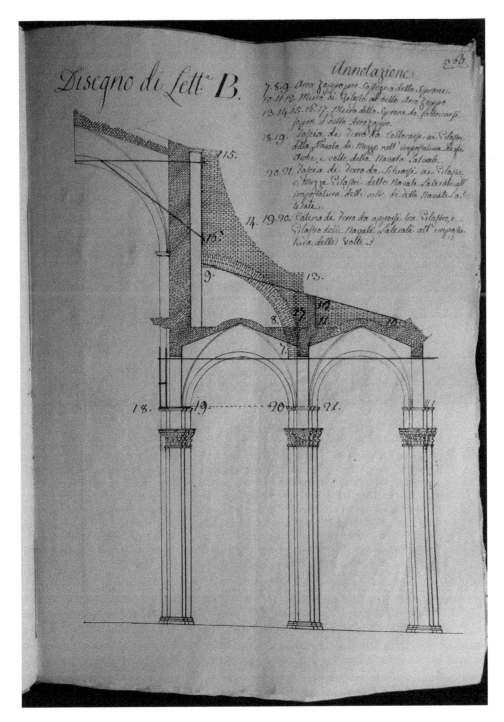

Figure 3.54 AOMS 1595 (932) 360r; drawing by Neri Zocchi and Giuseppe Salvetti (1 March 1800); flying buttresses and ties for the aisles of the apse.

Figure 3.55 AOMS 1597 (905ter) DSC_0953; drawing by Giovanni Gani; "Taglio, o alzato
trasversale di quella parte di fabbrica del Duomo di Siena, che rimane dietro
alla cupola, quale fu danneggiata dal terremoto del 26 mag[g]io 1798 dove si
dimostrano gli sproni e catene fatte apporre dai Sig[no]ri arcchitetti Giuseppe
Salvetti e Neri Zocchi, le catene fatte apporre dal Sig[no]re Ing[egne]re Bernardino
Fantastici e l'armature degli arconi rovinosi fatte da Maestro Domenico Vannini"
[Cut or transversal elevation of that part of the building of the cathedral of Siena
that remains behind the dome which was damaged by the earthquake of 26 May
1798 showing the spurs and ties affixed by the architects Giuseppe Salvetti and
Neri Zocchi, the ties affixed by the engineer Bernardino Fantastici and the frame-
work of the ruinous arches made by master Domenico Vannini].

Finally, the visible ties in the aisles were executed according to the original project (Figure
3.57) (Sembranti, 1997, 101–106; Gennari, 2005, 308–311). As regards the bay of the apse
divided into five aisles, the visible ties were always placed in just the bays adjacent to the
nave, while the paired ties at the keystones of the arches were extended to both arches on
each side. In this case, the anchors on the perimeter walls entailed problematic anchoring
through the splays of the windows.

The visual impact of both the ties at the keystones in the aisles and the anchors above the
capitals of the pillars was mitigated by continuing the bands of the pictorial apparatus on
them; the results obtained appear problematic, in particular the metal bands that anchor the
visible ties (Figure 3.58).

Figure 3.56 The festoon flying buttresses of the apse area emerging from the roofs of the aisles (a) and incorporated into the garrets (b).

The inspection of the garrets of the apse provided further information on the works to the vaults and the buttresses, which followed the earthquake of 1798. Their completion in 1801 was marked by an inscription painted just below the loggia of the drum, on the side adjacent to the apse.

In the garret of the nave, close to the façade of the baptistery, part of a web with distinct characteristics can be identified. With respect to the adjacent webs, it has a clearly discontinuous extrados due to the lack of coplanarity and continuity of the wall structure; it also has brick masonry which is differentiated by wider joints and the use of mortar with aggregates of bigger granulometry. This part of the web is most likely the only part of the medieval vaults that survived the earthquake of 1798 (Figure 3.59).

Figure 3.57 On the AOMS 1597 (905ter) DSC_0953 drawing are indicated the works carried out in different phases: the ties (green) inserted by Bernardino Fantastici
and the flying buttresses (pink) and ties (red and blue) created by Giuseppe
Salvetti and Neri Zocchi.

On the extrados of the diagonal arches, starting from the perimeter wall and for a stretch
of around 3 m, there is an inclined stone wall, 47 cm wide which reaches its minimum height
at the keystone and its maximum at the haunches of the arches. This wall may have been
connected to the stone elements that Fantastici had inserted into the diagonal arches. The
diagonal tympanum arches were created with it, with the intent of limiting the deformations
of the vaults during earthquakes, especially near the haunches. This method of stabilizing
the vaulted system would seem to be an alternative to dry material filling, the use of which
was rather limited here.

Finally, the existence of extrados longitudinal brick arches built on the vaults in two
parallel alignments is also worth mentioning. Two walls were built on these longitudinal
arches in order to obtain a higher garret in the central area, which was easier to access than
the one below where the roof heights were dictated by the lateral cornices. The groin vaults
were constructed with variable thicknesses: at the extrados of the lower zones there is a
clearly visible half-brick wall offset. In the garret, there are no visible metal, horizontal or
inclined fragments of the ties joining to the transversal arches, but they are partly visible
at the intrados.

Figure 3.58 Visible ties (red arrow) and ties that skirt the keystone of the transverse arches (green arrow) were inserted into the aisles of the apse area; the anchors to the semi-columns were created using visible metal bands and posts embedded just below the masonry surface (yellow).

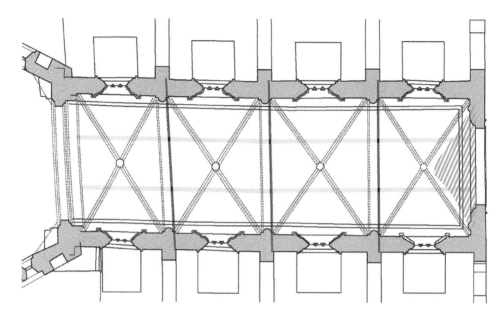

Figure 3.59 Plan of the nave of the apse area with the reconstruction works of the vaulted system following the damage caused by the earthquake of 1798; a small section of the original vaults still survives (purple lines) adjacent to the façade of the Baptistery; the stiffening walls were constructed above the diagonal ribs of the groin vaults (yellow); a double system of longitudinal arches (cyan), resting on the extrado3s of the vaults, supported the walls of an accessible garret positioned in the centre of the roof; metal ties (red) flank one side of the transverse arches.

The proposed works also envisaged that the apse roofs be supported by corbels inserted into the sides of the flying buttresses and diagonal rafters resting on the wall of the clerestory. This would have avoided, or at least limited, the weight of the roof on the vaulted system. The solution was significantly changed during the execution phase, the corbels in the buttresses were not used and were only created in the southern garret. So, in the southern aisle, on the sides of the flying buttresses the roof was supported by pillars placed on diaphragm arches. The roofs of the northern aisle sometimes rest on pillars which, flanked by flying buttresses, weigh on the webs of the groin vaults.

The stability of the surviving nave of the Duomo Nuovo was assessed and surveyed by experts, above all the part of the two open bays now considered a loggia. An initial report (1 March 1800) by Giuseppe Salvetti attributes the tilt of the pillar and the deformation of the arch between the two bays to events linked to the construction of the new cathedral, following which ties were inserted into the portico. Salvetti suggested erecting a wall between the two bays, with an archway in the middle and an oculus above it (Gennari, 2005, 318–322). The proposal was not followed, except for the construction of the foundations of the new wall structures planned. Consequently, a further suggestion made (in July 1803) by the engineer Cosimo Minucci proposed reinforcing walls for the pillar and the semi-pillar, recalling the precedent of the buttresses placed up against the sides of the aisles, as well as the replacement of the existing tie which was judged to be strongly compromised by evident corrosion.

Based on the analysis of the inclination of the structures, he considered the measures proposed in the previous opinion to be fairly ineffective. Lorenzo Turillazzi, roads supervisor, at the urging of the Governor of Siena, examined the earlier proposals (5 December 1803). He considered the use of buttresses to be always a loathsome solution in buildings of great importance; so instead, he suggested thickening the inner side of the pillar, so it was stylistically integrated and not a mere addition. However, later in the report, he made a different proposal, the aim of which was to eliminate the thrusts on the pillar by dismantling the brick vaults, which were to be rebuilt with lath and plaster. Turillazzi admitted that this solution was a "fallback", but it would have the advantage of not "deforming" the loggia. The reasoning clearly recognized the primary importance of achieving a coherent visual result, considering the true expression of the characteristics of the architecture and placing the importance of the actual constructive solidity of the loggia marked by the passage of time in the second place. Turillazzi also attributed the original deformation of the pillars to the construction of the Duomo Nuovo; in his opinion, they were subjected to the thrust of the vaults too early on, before the structure of wall was cohesive and fully efficient and the mortar fully solidified (Gennari, 2005, 356–362).

It was decided to ask the mathematician Massimiliano Ricca for a fourth opinion (29 December 1803), who expressed his thoughts on the previous reports, approving of those by Minucci and Turillazzi. Ricca attached Cosimo Minucci's report to his opinion, referring to the part where it mentioned the incompleteness of the Duomo Nuovo, believing that due to the "fatal plague that occurred at that time [in 1348] it remained imperfect [the construction]". A further opinion (18 April 1804) provided by the Royal Mathematician Pietro Ferroni added interesting considerations but did not hold back on expressing harsh criticism for the previous surveys that envisaged wall thickenings. He judged the solution proposed by Salvetti to be "vandalistic and barbaric" and that of Minucci unfit to preserve the "character of ancient architecture". This last solution taken up and reworked by Turillazzi was defined by Ferroni as "loathesome and disharmonious", "defective and bizarre" and even "monstrous" (Gennari, 2005, 362–369).

He believed that the new foundations of the wall additions could have worsened the inclination of the pillar and accelerated its fall. As proof of this he pointed to the behaviour of the building with respect to the test of time. Ferroni observed that six years had passed since the earthquake, but the condition of the loggia structures did not appear to have worsened. He also recalled that in 1780, Ximenes had considered the Torre del Mangia to be unsafe, and that in 1785, Salvetti considered the Orcia Bridge on the "Roman Royal Route" to be close to ruin. He opposed any intervention in the two buildings, smugly pointing out that they still stood "very firm, defying the ravages of time."

Referring to Turillazzi's opinion, Ferroni proposed dismantling the brick groin vaults and replacing them with light lath and plaster vaults, as well as affixing ties to secure the out of plomb pillars. The project was approved in May that year, but the brick vaults were not demolished, while reinforcements were probably made using ties.

The interventions following the earthquake of 1798 led to the completion of a profound constructive reorganization of the cathedral, which was started in the 17th century and again dictated by the critical conditions the earthquakes had gradually caused. Thus, over time, all the structural protections that had been rejected by the medieval architects were used. Buttressing systems were used in the basilican body, placing brick walls on one side and wooden supports on the other; the transverse arches of the nave were supported by ties placed at the level of the keystones. This solution was re-proposed, with some variations, in the nave of the apse, where massive flying buttresses were constructed to help offset the thrusts of the vaults; in the apse, concealed and visible ties were also added in the aisles.

Although there is no documentary proof, the reinforcement works carried out at the base of the drum probably date back to this same period.

In this case, the particular construction context of the drum and the transept should be considered first. The cathedral's position on a slope means that the walls of the transept wing on Piazza Duomo are lower in height than those of the other wing adjacent to Via Fusari.

This aspect should be considered in view of the relationship between the crossing and the transept. Four pillars of the hexagon, placed on the south-east and north-west sides, have similar transversal counterthrust systems, made up of diaphragm arches extending for one bay and attached to walls. This is also due to the fact that in each of the wings of the transept, a span was transformed into a chapel by infilling its arches; this is how the chapels of St John the Baptist (1482) and the Madonna del Voto or Chigi were created (1559–1562) (Carli, 1979, 109, 130–132). The two remaining pillars, on the side of the apse, each have transversal reinforcements divided into three diaphragm arches abutting the perimeter walls. The decentralized position of the crossing with respect to the transept should still be pointed out as the two major arches, placed transversely, are placed in very different construction contexts. The one near the façade of the cathedral is flanked by diaphragm arches that abut the bell tower on one side and on the other, the side of the Piccolomini library placed longitudinally; this arch is therefore protected by very rigid structures. The situation of the arch on the opposite side, towards the apse, is different and not only because it is flanked on each side by three diaphragm arches abutting perimeter walls placed transversely; in fact, this arch is followed by two spans divided into seven and five naves, while on the façade side there is a basilican body divided into three aisles.

At the end of all of this, torsional effects could have been created in the pilaster-drum-dome system during the earthquakes, resulting in damage to the adjacent vaults. Evidence of this could be the structural reinforcements deemed necessary by the builders in the north arm of the transept.

In fact, three counterthrust systems were positioned near the corners of the hexagon of the drum.

A brick strut with a 62 × 62 cm section orthogonally abutted the north-west wall, between the trifora and the niche. The strut, interlocked with the drum with a clear recess and fitted with a head stone ashlar, would suggest that it was initially isolated and later incorporated into the masonry, which in fact is not aligned with the strut (Figure 3.60). A test was carried out at the foot of the brick strut. Under the screed, the strut continues in the filling of the vaults. The test, carried out to a depth of 60 cm, did not reach the support level of the masonry strut. However, as highlighted and as could be expected, the strut climbs over the web below and is most likely placed near the outer wall, resting on the transversal diaphragm arch that divides the first bay of the aisle and the adjacent triangular bay of the transept, while the wall beneath the strut was built on the filling of the vaults (Giorgi & Matracchi, 2010).

At the north corner of the hexagonal drum, a sort of flying buttress incorporated into the garret was created. The flying buttress is divided into three archways and has two intermediate pillars resting on the diaphragm arch below. The two outermost openings have an archivolt and are limited in width, the internal one is a much higher opening which was probably enlarged later to connect and make the different areas of the garret more accessible. This latter opening has limited the structural continuity between the flying buttress and the drum, compromising its effectiveness.

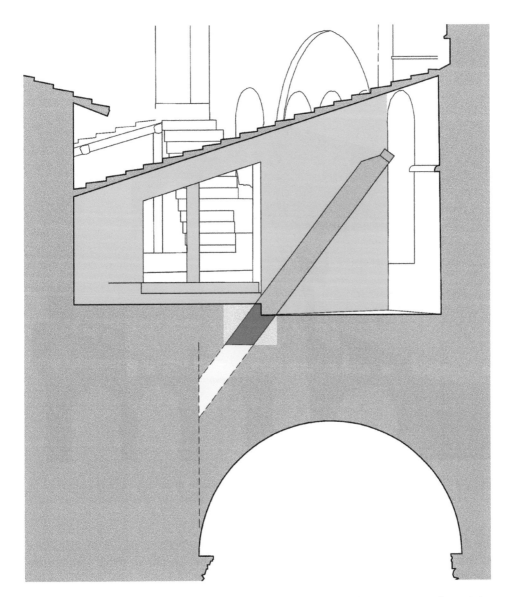

Figure 3.60 Detail of the buttress of the north-west side of the drum; it was confirmed that the masonry support (pink), incorporated into a wall (green), continues into the sides of the vaults (red) and probably extends (yellow) until reaching the side of the nave.

The third counteracting structure is aligned with the arch of the crossing at the apse. It is a buttress confined to the garret and built on the diaphragm arch adjacent to the drum (Figure 3.61).

The three buttressing structures are always placed on the diaphragm arches of the bays adjacent to the drum. In the first two cases, they directly abut the perimeter walls of the transept;

Figure 3.61 Buttressing system for the part of the hexagonal drum opposite the square; the structures are placed to in counteract (a) the west, (b) north-west and (c) north corners.

in the last case, the counteraction wall is interrupted at a pillar, but here, the contribution of the underlying diaphragm arch divided into three bays must be considered, which transmits the thrust to the transept wall.

As regards the north wing of the transept, in the bay adjacent to the chapel of San Giovanni Battista, it should also be noted that the vault required a diagonal rib to be strengthened with transverse metal brackets anchored to an extrados wooden element.

The archive documents certify the purchase of metal elements to reinforce the vaults (Sembranti, 1997, 79 note 52, 84 note 66, 91). Today, in several bays, transverse brackets can be observed in the ribs of some vaults combined with wooden elements placed on the extrados. The wooden elements are connected to the diagonal arches with a single central bracket or two brackets at the ends. On the intrados, the brackets were shaped to fit the profile of the ribs. In the north aisle of the apse, a long section of a diagonal rib has been strengthened using four distinct wooden elements at the extrados; it can be seen that on this groin vault, there are two pillars supporting the roof.

3.5 The transformations of the dome

The decision to renew the lantern of the dome was made in the 1660s (Butzek ed., 2006). The drawings for the design, despite their schematic nature, represent some key aspects of its actual condition. The internal dome consistently has two different thicknesses and metal brackets on the extrados to support the lightweight outer dome, probably made of a wooden structure covered with lead sheets. The bulbous dome emerging from the double dome and surmounted by a ball is of moderate size. The parts below were also drawn, highlighting the drum, with the blind inner loggia and the level of the trifora and the pillars of the crossing. One of the designs developed included a metal structure anchored to the masonry dome. The solution that prevailed, albeit with some variations in the study phase, was based on a masonry structure consisting of pillars combined with columns. There is a clear compromise in the architectural characteristics, which used triforas as an element of continuity with the past, alongside styles more consistent with the construction era. Particular attention was given to the connection between the lantern and the dome; in fact, the hollow space between the two domes housed the base of the lantern, with elliptical openings and placed on six sturdy brick ribs connected at the top by arches (Figure 3.62).

The prudence of adding ribs at the extrados of the medieval dome was well justified by the fact that the lantern is a sort of tower placed on the dome; this lantern, between the springer on the internal entablature and the summit, is 9.00 m high, while the lantern with the crowning ball is over 12.00 m high.

The ribs, one-brick thick (the thickness varies from 30 to 37 cm), have a variable width which gradually reduces from the base (1.20/1.30 m) to the top (0.75/1.00 m). The decision to create six ribs was linked to the fact that each have a support oriented to a pillar of the crossing. It should be said that the bases of the ribs are not placed in correspondence to the pillars of the crossing; structural continuity between the ribs and the pillars is obtained through the masonry corbel (Figure 3.63). Considering the reciprocal vertical position between the pillars and the bearing surface of the ribs, highlighted by the sections, the latter overhang by 0.75/0.90 m. These ribs were built, without interlocking, on the extrados of the dome which was settled with fillings of mortar and adhering masonry; this was particularly necessary in the wall offset area at the extrados where the thickness of the dome reduces from one-brick-and-a-half to one-brick thick. Here, at the extrados, the dome seems to have an inflection at the estrados, which is actually real in the case of the ribs, but it is due to the characteristics of the bearing surface and not displacement.

On 17 October 1890, during maintenance work on the roofs, a fire broke out, and in just a few hours, it had destroyed a large part of the roofing of the basilican body and the outer wooden dome covered with lead. On 27 July 1891, Giuseppe Partini, appointed by the Opera del Duomo, submitted the project to reconstruct the roofs and outer dome for approval (Galli, 2006, 357–363, 398–405, figs. 53–64).

The main aim of the proposal was to avoid the use of wood, replacing it with metal structures. As for the dome, six brick ribs with a semicircular section were planned, placed alongside metal reticular structures to support the roof structure constructed with an iron frame, segmental brick vaults and a screed finish. The committee appointed by the Ministry of Public Education, made up of Luca Beltrami, Luigi Del Moro and Giuseppe Sacconi, despite understanding the purposes of the project, asked to adopt construction solutions that were closer to the characteristics of the dome. Partini, returning to an original idea of

Figure 3.62 Views of the hollow space between the two domes.

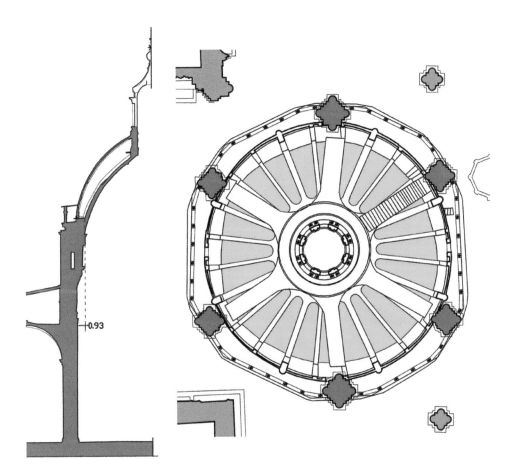

Figure 3.63 Section that shows the overhanging support of the dome with respect to the pillar below (a); (b) plan with the extrados of the internal dome (green) in rela-tion to the pillars of the crossing (red).

his, obtained consent from the committee, opting for a brick dome covered with lead sheets, which were deemed more appropriate than the alternative of copper (Marini, 1981, 186).

Many decisions on the construction aspects must have been taken during the work on the dome, probably completed by Partini between 1891 and 1893 (Galli, 2006, 362). The choice of the brick dome and the decision not to use metal structures implied a radical rethinking of the hooping devices and the connection with the medieval dome.

In any case, from a geometric point of view, the new dome is entirely different from the previous one, which had clear planimetric and profile irregularities (Figure 3.64). In fact, the plan of the outer dome deviates slightly from a circumference, and its transversal section is a "three-fifth" pointed curvature (the dome's radius is equal to 3/5 of the base diameter).

Despite similar geometric differences, the ribs of the lantern affected the arrangement of those of the outer dome, which became eight during execution and not six as indicated in the project. Between each pair of semicircular ribs smaller visible ribs were added at the

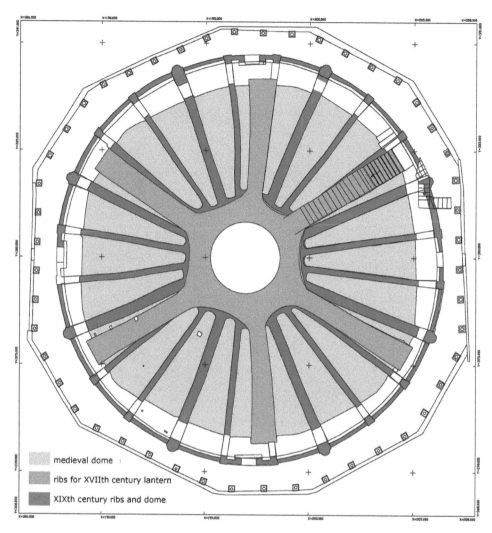

Figure 3.64 Plan of the extrados of the internal dome (light grey) with the complex ribbing system; first the larger ones were added to support the lantern (dark grey), and, following the fire of 1890, the external dome was reconstructed with the thinner ribbing (red).

intrados; in correspondence to all the ribs of the outer dome, ribs were constructed on the extrados of the medieval dome, which are connected at the top by small arches, replicating the solution of the ribs of the lantern. This system of ribs at the extrados of the medieval dome was only fully achieved between the ribs of the lantern on the sides facing the apse and the façade, where there are bigger spaces. This congruity with the ribs of the lantern gave rise to the positioning of the rest of the ribs of the new dome. As a result, the positioning of these latter appears to be rather casual; in fact, there is a major rib accompanied, in varying positions, by two minor ribs and, in some cases, the minor ribs were either not built or they were partly constructed so that they did not overlap with the ribs of the lantern. The 19th-century ribs were interlocked with the main dome in five points, placed at increasing intervals from the bottom to the top. The 19th-century ribs have marked inflections as they follow the irregular curvature due to the wall offset at the extrados of the main dome. The flat-jack tests showed that the ribs are not subject to compression action, and in some cases, they are actually subject to traction (Bartoli et al., 2013, 24–33).

The dense ribbing system of the dome, made up of eight major ribs, semi-circular on the outside and with an internal overhang of 45×45 cm, alternating with 16 minor ribs at the intrados of 17×32 cm, stems from the intention to create a thin dome just 15 cm thick. The hoopings connecting the dome and ribbing was created using different devices. Two travertine ties, with a 22×20 cm section placed at a height of 2.25 and 4.10 m from the springer level, form a polygon of 24 sides; the stone elements on each side are probably connected by clamps in the internal parts of the ribs. In addition, there were two metal ties alternating with the stone ones and anchored at the extrados of the dome by through bolts; the dome was therefore hooped up to a height of 5.85 m from the springer, more than half of its vertical development (Figure 3.65).

The outer dome has 24 separate supports, each corresponding to a rib, made of special pieces of inverted T-shaped travertine, which on the outside assumes a toroidal shape in the case of the major ribs. These stone elements also provide the springers for the structures that complete the support of the dome: segmental arches and jack arches below. At the doors providing access to the hollow space between the two domes, only the jack arch that creates a sort of step was made (Figure 3.66).

Between the ribs added at the extrados of the main dome and those of the new dome, a connection was created using travertine horizontal bars, specifically placed in correspondence to the hooping of the outer dome. In particular, three bars arranged on the same vertical plane are pinned on the eight main ribs, at a distance of around 1.30 and 1.65 m from each other going up, two of which correspond to the height of the stone ties, one at the level of the metal tie-rod placed at the highest height (Figure 3.67). The minor ribs are connected to the outer dome by a single travertine bar, placed at the same height as the stone tie placed lower down. In just one case is the connector placed between a minor rib and the ribbing of the lantern (Figure 3.68). The travertine bars have a constant width of 20 cm, while the height is 20 cm for those on the first level and 27.5 cm for the rest.

The horizontal stone connectors have several cracks that can be attributed to tensile forces, which acted close to the intrados or extrados. In some cases, with respect to the mechanical characteristics of travertine, the connections with the masonry have less tensile efficiency; in this case there are intact stone elements and cracks in the masonry adjacent to the joint (Figure 3.69). In more detail, in correspondence to the major ribs combined with three levels of travertine bars, there are three ribs which all have intact stone elements, three ribs with horizontal stones only damaged at the first level, and two ribs with an intermediate intact stone. Only five of the sixteen travertine bars of the thin ribs are intact (Bartoli et al., 2013, 18–21).

As the support of the external dome is circular, it cannot be congruent with the irregular dodecagon drum below. Thus, the pairs of major ribs on the sides facing the apse and the façade, from which the dome was set, are perfectly centred on the wall below, while two ribs almost completely rest on the paired-column loggia, right at the transverse lintel. It should be noted that between the support of the rib and the extrados of the loggia vaults, there is a compact wall mass of about 1.82 m high. Therefore, the considerable masonry mass below the ribs helps to distribute the loads over the extensive area of the vaulted system below.

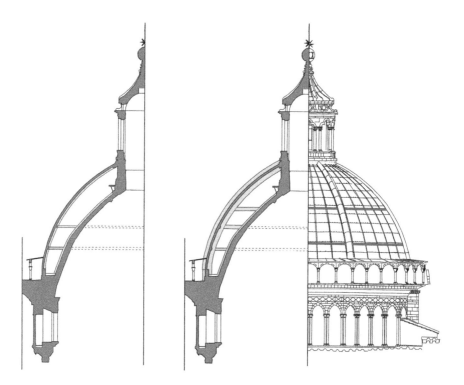

Figure 3.65 Sections and elevation of the dome; (a) corresponding to the thinner ribs, there is a level of stone ties (green); (b) the larger ribs of the external dome (cyan) are counteracted by three levels of stone ties (green); the two lower ties are connected, in the external dome, to stone hooping (yellow); (c) two metal rings placed on the extrados of the external dome were added (red).

Figure 3.66 Detail of the upper loggia and the base of the external dome.

Figure 3.67 Hollow gap between the medieval dome and the 19th-century one; the circumferential stone ties (red), the transverse stone ties (yellow) and the larger (green) and smaller (cyan) ribs of the external dome are highlighted.

At the top of the masonry mass above the paired-column loggia, some superficial cracks have formed right at the edge and parallel to the perimeter of the 19th-century dome, which could be due to slight subsidence caused by the weight of the dome itself.

Many of the transverse lintels of the paired-column loggia needed to be reinforced with metal brackets. Leaving aside the particular solution of the lintels corresponding to the single-light windows, the cracks always formed near the drum wall; this is probably due to the different constraint conditions at the ends of the beams, where a fixed joint in the side of the wall and a hinge on the support above the capitals of the paired columns can be considered. This particular condition makes it plausible that the cracks in the lintels, rather than being caused by the construction of the external brick dome, are due to the numerous earthquakes that struck Siena over time. In a condition of dynamic stress, it is easy to imagine strong states of tension close to the joints of the transversal stone lintels.

3.6 The cathedral as a palimpsest of structures

Siena cathedral is an extraordinary example of architectural complexity understood as a concrete act of building, the result of several construction phases linked to countless specific conditions.

It has unique genetic conditions. The orientation of the cathedral was dictated by the earlier cathedral, which for a certain period of time necessarily had to exist alongside the construction works to ensure continuity of worship. This placed the apse and transept on a slope and in a geologically distinct area given that it is at the edge of the fault line of a conglomerate bank.

The works continued with the construction of the basilican body and later the domed space, with quite a few geometric adaptations. At the same time, it became necessary to make changes to the crypt and the apse by introducing two pillars into this space. This led to the completion of a cathedral, which turned out to be intermediate.

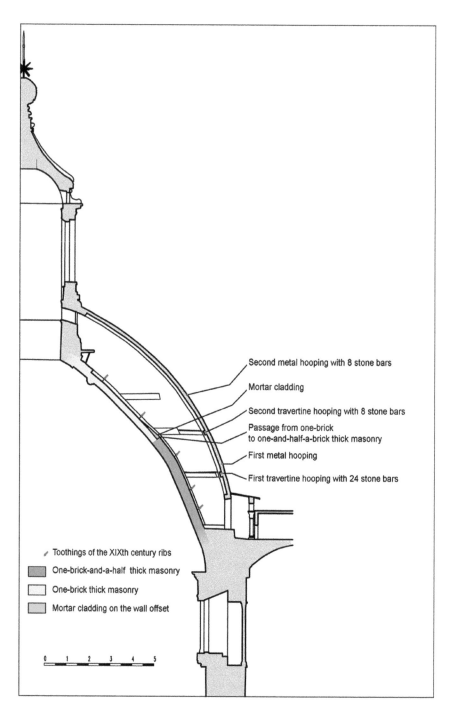

Second metal hooping with 8 stone bars

Mortar cladding

Second travertine hooping with 8 stone bars

Passage from one-brick
to one-and-half-a-brick thick masonry

First metal hooping

First travertine hooping with 24 stone bars

Toothings of the XIXth century ribs

One-brick-and-a-half thick masonry

One-brick thick masonry

Mortar cladding on the wall offset

0   1   2   3   4   5

Figure 3.68 Section of the dome indicating the construction elements and points of connection between the thin ribs and the dome.

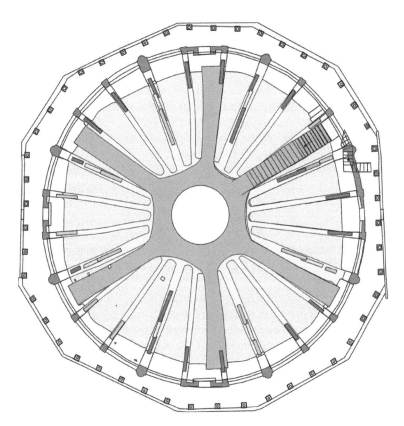

Figure 3.69 Plan of the dome indicating the stone ties connecting the external dome to the ribbing; the cracked ties are shown in red and the intact ones in green.

In fact, first the basilican body was extended by one bay, and the present-day façade was built. It was then decided to build a new baptistery taking advantage of the slope behind the apse; a new and larger apse with a basilican plan was built on this same side, above the baptistery. Naturally, this involved dismantling not only the apse but also the transept of the intermediate cathedral.

The decision to build the Duomo Nuovo was made while the apse was being constructed. This new cathedral was extremely ambitious and considerably larger than the existing one, which would have become the transept of the new construction, or it would have been divided into three sections, of which the apse and the basilican body would have become independent churches, and the crossing area would have been integrated into the new construction. But the basilican body part of the Duomo Nuovo developed such serious structural problems that it had to be dismantled and may have partly collapsed. This led to the completion of the baptistery project.

In addition to the genetic factors, the conditions of the orographic context and the results of construction choices that turned out to be inadequate, there were unpredictable factors. The repeated action of earthquakes and the consequent damage between the 17th and 19th centuries led to a complete rethinking of the cathedral's construction aspects. All the protections that the medieval creators had not wanted to use, and paid the consequences for in the Duomo Nuovo, were gradually added. In the basilican body, masonry buttresses were added on one side of the nave and wooden struts on the other. Moreover, ties were introduced at the level of the intrados of the transversal arches. In the apse symmetrical festoon buttresses were constructed and ties were inserted in the nave and aisles (Figure 3.70).

There were also changes linked to accidental events. During the maintenance works on the lead roofs of the basilican body, carelessness on the part of the builders allowed a fire to break out that also engulfed the external wooden dome, and as a result, the present-day brick dome was built. But, even earlier, in the 1660s, the desire to innovate the medieval dome led to the construction of the lantern that still exists today.

An extreme but effective summary of the cathedral's tumultuous construction history and elements of the context that influenced it can be seen by comparing some cross sections. This provides an overview of the great complexity and variety of the construction aspects.

Starting with the façade, the section in the first two pillars of the basilican body shows that the thrusts of the vaults of the nave are counteracted by a brick buttress on one side and wooden supports on the opposite side; there is also a tie at the top of the transverse arch with long lateral anchoring posts.

The next section shows the arch of the crossing at the end of the nave. This arch is flanked by the diaphragm arches of the aisles, which are extremely rigid because on one side they are buttressed by the bell tower and on the other by a wall of the Piccolomini Library. It should be pointed out that these very rigid protections are located just below the capitals of the

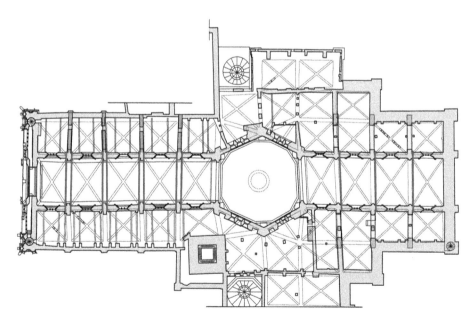

Figure 3.70 Plan at the garret level of the cathedral; it highlights the buttresses (blue), the ties (red) flanking the transverse arches and the wooden struts (green) in the basilican body, and in the apse area the flying buttresses (yellow) and the ties (red) placed on one side of the transverse arches.

central arch. This particular arrangement in fact explains the slight inclination of the pilasters towards the inside and the slight downward sliding of the keystone of the central arch. It is quite likely that these modest deformations were caused by earthquakes, while the remaining parts were firmly protected by the massive flanking structures.

It is interesting to compare this section with that of the opposite side of the crossing, where the structural arrangement highlighted by the section is significantly different. Here the central arch is lower in height, so its haunches correspond to the diaphragm arches, there being three on each side in this case abutting the perimeter walls of the transept. It should be added that here the pillars of the crossing continue into the crypt; moreover, on the Via dei Fusari side the perimeter wall is considerably higher than on the opposite Piazza Duomo side. On the Via dei Fusari side only, a buttress was constructed in the garret above the first diaphragm arch. The cracks of the central arch are different to those observed in the previous section. A slight bell-shaped crack has formed in the central part of the arch; it is likely that here, due to seismic action, the supporting construction system is less rigid, and the entire archway has become slightly deformed. Moreover, the possible torsional effects produced on this side by the dome-drum system mentioned above should not be overlooked. The cracks in the two transverse arches also extend to the archivolt openings of the drum above.

The structures flanking the arch towards the apse have also undoubtedly demonstrated remarkable effectiveness, taking into account that the entire vaulted system of the adjacent nave of the apse suffered very serious damage after the 1798 earthquake and was rebuilt. A close inspection in particular of the crack in the arch towards the apse might be useful.

Finite element modelling was carried out in order to produce a preliminary evaluation of the effects of seismic actions on the cathedral structures. As for the two major arches of the crossing, the introduction of ties has proven not to be very effective (Bartoli et al., 2013, 116–126).

Returning to the analysis of the cross sections, the one at the centre of the dome shows the base of the drum supported by diaphragm arches, which in turn abut sizable walls. A wall close to the drum has been constructed in just one garret.

The first cross section of the apse has a five-aisle layout. The nave is flanked by festoon flying buttresses that rest on pillars and continue with buttresses up to the perimeter walls, where there are transverse walls. Ties have been placed in the nave at the keystones of the transverse arches and are anchored to the sides with a system of bars placed in a triangle. In the aisles the ties are placed at two levels, at the springer and at the keystones of the arches. All the pillars continue in the underground spaces and the structures are higher on the Via dei Fusari side.

In the next section the structure is divided into three aisles, the pillars are placed at the corner of the walls that flank the apse of the baptistery. The festoon buttresses directly abut the massive transversal walls; but on the Via dei Fusari side, the buttress rests on an arch of the underground rooms. The tie reinforcement system of the naves follows that of the previous section.

Finally, the section that intercepts the last pair of pillars shows the connection with the central pillars of the baptistery below. The structure reflects the double level that the cathedral has here, with the apse fully overlapping the baptistery. The flying buttresses rest on the perimeter walls arranged longitudinally. The aisles were given the same reinforcement systems used in the apse.

The sequence of sections (Figure 3.71), which is incomplete as others could have been added, effectively shows the variety of structures used to construct the cathedral and the

Figure 3.71 Comparison between the transverse sections (red), executed in correspondence to the pillars, with the ties highlighted.

absence of symmetry even in the individual sections, except for the final section of the apse.
The changing nature of the structural elements also includes local aspects that are no less
important, such as the deformations and consolidation of the ribbing of the vaulted systems
of the aisles, or the particular construction solutions of the loggia with its paired columns,

with vaults resting on transverse stone lintels, which, in correspondence to some single-light windows, are in turn connected to vertical stone tie-rods. Not to mention the medieval dome which, leaving aside its external form, has its base integrated with the first level of the loggias of the drum. The dome has a dense network of 17th- and 19th-century ribbing at the extrados, on which the lantern rests and horizontal stone bars are anchored to the outer dome. However, the two ribs of the lantern that were checked are not subjected to compression forces. The lantern, therefore, leaving aside the structural alignment created at the base, rests for the most part, or even entirely, on the medieval dome. The outer dome, in turn, can only rely to a limited extent on the stone horizontal connections, for the most part broken into two by cracks. But this did not result in any cracking on the external dome, which is evidently effectively protected by the stone and metal hoopings. The discoveries made in the dome revealed that at times, the ineffectiveness of the construction systems put in place did not affect the stability of the ancient structures. This shows how ancient buildings have resistance resources which, from today's modern perspective, are sometimes difficult to understand and appreciate.

Chapter 4

Florence cathedral

4.1 The start of a complex building site: the renewal of the ancient cathedral of Santa Reparata and the bell tower started by Giotto

The present-day Florentine cathedral of Santa Maria del Fiore (Figure 4.1) is the final product of a construction phase that started in 1296 to give the city a cathedral worthy of the political and economic importance Florence had acquired at the time. It was therefore decided to enlarge the square between the church of Santa Reparata, the cathedral at that time, and the Baptistery of St John, setting back the façade, raising its level and expanding its dimensions. The design of the new building, which was only dedicated to the Virgin Mary during the works and no longer to Saint Reparata, was continuously adapted, even with extensive changes, and only reached a first conclusion one and a half centuries later with the construction of the lantern of the dome, but it was continuously altered in subsequent centuries until assuming its current form at the end of the 19th century.

Santa Reparata, founded in the 6th century (Belli, 1994, 26–27) and thereafter transformed and expanded several times, in the 13th century must have appeared as a modest cathedral alongside the size and opulence of the Baptistery (Figure 4.2), positioned a short distance from its façade which may have even had a portico. The façade's position has been confirmed by georadar imaging, which places it close to the steps of the parvis of the present-day façade of Santa Maria del Fiore (Morelli, 2006, 226–227).

The remains that can be seen today underneath the cathedral show that Santa Reparata had a mosaic floor at a height just above the Roman levels found in the room below the Baptistery (Marino et al., 1996) (Figure 4.3). Other Roman levels were found just below the floor of Santa Reparata. The significant coincidence between the Roman level and the floor of Santa Reparata, due to its distant founding, helped to accentuate the grandeur of the baptistery, which was built at a higher level, most likely from the 10th century onwards, establishing the current height of the cathedral square.

However, when around the end of the 13th century there were calls for the renewal of Santa Reparata, the floor level of that church was the one established between the 11th and 12th centuries (Morozzi et al., 1974, 11, 20–21), positioned above the mosaic level but in any case, around 50 cm lower than the Baptistery (Figure 4.4).

From 1293 to 1296, the city council continuously set aside increasing amounts for work on Santa Reparata. According to a document dated 10 February 1296, Pope Boniface VIII acknowledged the wish to construct a more magnificent building. The church was under construction the following year (Riccetti, 2001, 210–216).

Figure 4.1 The Cathedral of Santa Maria del Fiore and Giotto's Bell Tower.

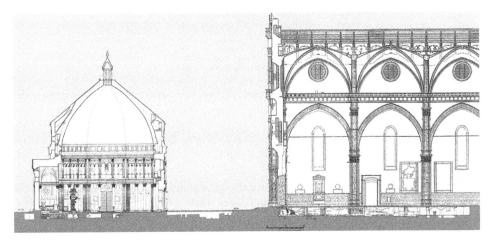

Figure 4.2 Section of the nave of Santa Maria del Fiore and the Baptistery, with the archaeological areas below and the cemetery area below the parvis.

The lack of more detailed documentation has made it difficult to interpret this start to the construction of a new Santa Reparata.

Some significant information was obtained during the work to create the present-day 19th-century façade of Santa Maria del Fiore. Emilio De Fabris, who produced the winning design,

Figure 4.3 View of the excavations of the ancient cathedral of Santa Reparata below the aisles of Santa Maria del Fiore.

confirmed the extent of the foundations by carrying out specific tests in several points, verifying the depth of 5.80 m from the road surface "up to the gravelly layer" (Cavallucci, 1881, append. I, 11).

Some traces found in the elevation led De Fabris and Luigi del Moro to speculate that the walls of the façade had been thickened (Cavallucci, 1881, 115 append. I; Del Moro, 1895, 14–).

Studies on Santa Reparata were significantly boosted following the excavation campaign carried out between 1965 and 1974, which also brought to light the foundation structures of the façade of Santa Maria del Fiore (Morozzi et al., 1974; Morozzi, 1987). As regards the latter, however, it was Luigi Marino who made a fundamental contribution, conducted on the basis of a specially executed survey (Marino, 1988). Some key aspects that emerged during this work concern the complex constructive genesis of the façade foundations, characterized by wall additions that increased both the thickness and the overall width.

A further study of the constructive aspects of the façade of Santa Maria del Fiore, conducted by Nicolò Masturzo, highlighted the initial focus on producing a façade that divided the three aisles body of the ancient cathedral (Masturzo, 2006, 210–215). Although Santa

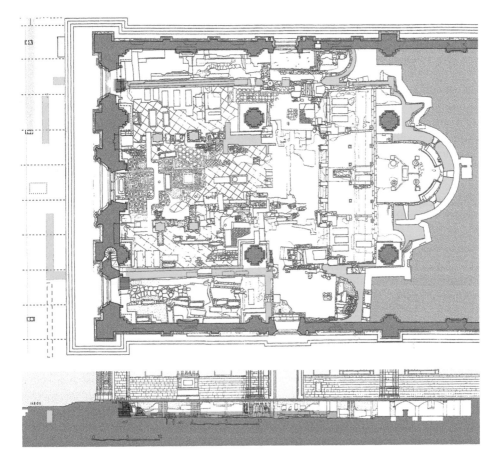

Figure 4.4 (a) Plan of the aisles with the remains of Santa Reparata (green); the position of the façade extending beyond the steps of the parvis has been highlighted. (b) Detail of the longitudinal section with the surviving fragments of the aisle pillars (green) and the level of the floor of Santa Reparata in the 11th and 13th centuries (orange); the foundations of a new façade for Santa Reparata, which was then not built, are indicated (red).

Reparata would have maintained the same overall width with this new lay-out, its façade would have been set back with respect to the Baptistery, and the floors would have been significantly higher. It clearly marked a significant restructuring of the ancient Santa Reparata, to be placed with the initiatives of 1296, which would have led to the complete renovation of the aisles (Giorgi & Matracchi, 2006a, 102–103).

In the subsequent phase, which led to the current dimensions of the façade, the previously laid foundations were thickened; moreover, they were extended beyond the basilican body of Santa Reparata. This structure is mainly made up of pit foundations connected by arches sometimes built directly on ground that was fashioned for this purpose (Marino, 1988, 17–20; Masturzo, 2006, 215–216).

With this last phase, the choice to permanently abandon the ancient cathedral of Santa Reparata could be considered complete. Nevertheless, the lack of documents makes it problematic to understand when this highly important change of design and the subsequent construction of the sides of the present-day cathedral occurred.

According to Giovanni Villani (1991, cap. 193), works resumed in 1331 after a long period of inactivity. In fact, earlier there were only reports of the acquisition of marble from Carrara and an increase in builders in 1318, while in 1331 the city council promoted work on the cathedral by setting aside more money (Guasti, 1887, 30–33).

The appointment of Giotto in 1334 as head of the cathedral construction site, as well as of the walls and fortifications of the city of Florence, following the city council's wishes, led to a turning point in the works. The foundations of the bell tower were laid immediately and completed with great speed between July and September of that same year (Figure 4.5).

Considering how slowly the construction of the cathedral had proceeded for some time, this choice may seem radical as it was extremely demanding in both organizational and economic terms. It has been speculated that Giotto had wanted to honour the city council's appointment by starting a new architectural work (Guasti, 1887, XLV–XLVI). The close connections between the cathedral and the bell tower should be pointed out, nonetheless.

The bell tower has a square plan with sides measuring twenty-five braccia (14,60 m) with semi-octagonal corners and is positioned a short distance from the side of the cathedral, ten braccia. According to tests carried out in 1639 by Gherardo Silvani, in the area between the two buildings, the foundations of the bell tower are much deeper than those of the corner of the cathedral façade. The connection, by means of an arch, between the foundations of the two buildings was the work of Silvani (Giorgi, 2017, 19–20).

Such close proximity between the bell tower and the cathedral is also linked to the connection they sought to create between the two buildings. An entrance door was created on the side of the bell tower facing the side of the cathedral to provide access to stairs that within the wall thickness leads to the upper levels until reaching the rooms on the first and second floors of the bell tower (Figures 4.6 and 4.7). From the level of the cathedral square, another staircase led independently all the way up to the top of the bell tower. This differentiated and extremely complex system of stairs allowed privileged access from the cathedral to two rooms of the bell tower; both routes could be accessed from the room on the second floor. However, this system of stairs was only able to develop into the current arrangement after the walls of the bell tower had been doubled, which occurred during the construction phase linked to Andrea Pisano (1337–1349) with the addition of an internal wall (Rocchi, 1996a) (Figure 4.8).

As regards the room on the ground floor, it has only recently been possible to confirm that the entrance door was sealed by a grate that slotted into a narrow space created especially in the walls, close to the inner side (Figure 4.9). A similar defence system for the room suggests that this room was intended to protect the precious furnishings of the ancient Santa Reparata, the basilican body of which had been partially dismantled for decades by this point and was in fact part of the construction site; indeed, this meant that the most valuable objects were highly vulnerable as far as their safety was concerned. A room in which to store the treasures of Santa Reparata had therefore been built on the ground floor of the bell tower.

The direct connection between the bell tower and the side of Santa Maria del Fiore, through a passageway over an arch described in a document dated 1397 (Guasti, 1887, doc. 408), was completed by an entrance created in the adjacent side of the cathedral, in the space between a mullioned window and a pilaster.

It should also be added that the rooms of the bell tower were used to hold the meetings of the canons of Santa Reparata. The bell tower conceived as a habitable tower is confirmed

Figure 4.5 Giotto's Bell Tower; east side, detail of the base and view of the walkway on corbels between the first and second mullioned windows.

by the toilets identified within the wall thickness, at the height of each room (Figure 4.10). On the other hand, the dismantling of the ancient Santa Reparata must have also affected the additional rooms used by the canons (information on the bell tower, pertaining to the functions of the rooms, was newly acquired during a study campaign promoted by the Opera del Duomo di Firenze from 2017–2018).

Figure 4.6 (a) Walled door created in the side of the cathedral between the mullioned window and the external pilaster. (b) Bell tower door.

The intention and high complexity of access to the bell tower from the cathedral confirm that the room on the ground floor of the bell tower was likely under construction at the same time as the side of the cathedral, or that these parts belonged to the same phase, in which they addressed aspects that simultaneously involved the ancient Santa Reparata, the cathedral and the bell tower under construction.

This period saw the start of a phase in which greater resources were allocated to the building of the bell tower, carried out from 1351 under the direction of Francesco Talenti, defined in the documents as "principalem magistrum".

It is interesting to consider the chronology of some documentary claims. On 5 February 1356 (modern-style dating, also used thereafter), work on the bell tower was progressing well (Guasti, 1887, LI, 66, 86).

On 12 May 1357, it was suggested that the city council increase the amounts being put aside, envisaging the completion of work on the bell tower in four years; it was estimated that if work were to continue with the funds available, it would take twenty years, moreover incurring considerably higher costs. In this phase of works, they reached the level of the large trifore.

Just one month later, in June 1357, a sudden change in the management of the construction site was recorded. It was decided (14 June) to drastically reduce the number of builders working on the bell tower and to concentrate most of their efforts in the cathedral (15 June), where work immediately started on the construction of the pillars of the bays, starting with the laying of the foundations (Guasti, 1887, LVII, 90–93).

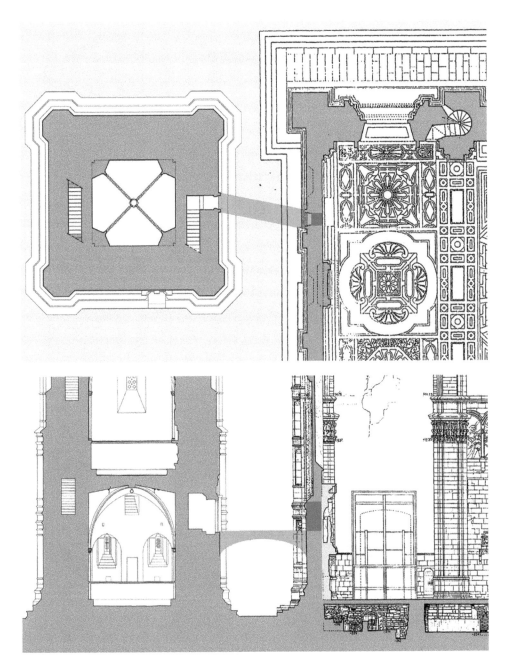

Figure 4.7 Plan and section with the connection (orange) between the side of the cathedral under construction and the bell tower; (red) walled door in the side of the cathedral.

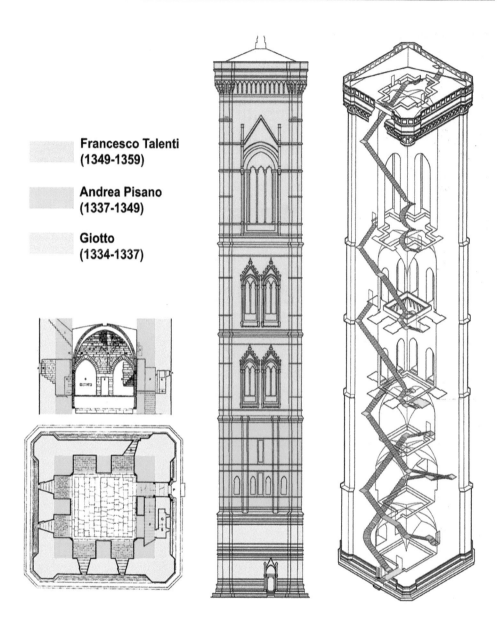

**Francesco Talenti
(1349-1359)**

**Andrea Pisano
(1337-1349)**

**Giotto
(1334-1337)**

Figure 4.8 The construction phases of Giotto's Bell Tower, the initial part by Giotto (yellow), the contributions made by Andrea Pisano (green) and Francesco Talenti (cyan). (right) The complex system of stairs that cross each other (green) at the levels of the three superimposed rooms, while in the top part by Talenti spiral staircases alternate with straight flights.

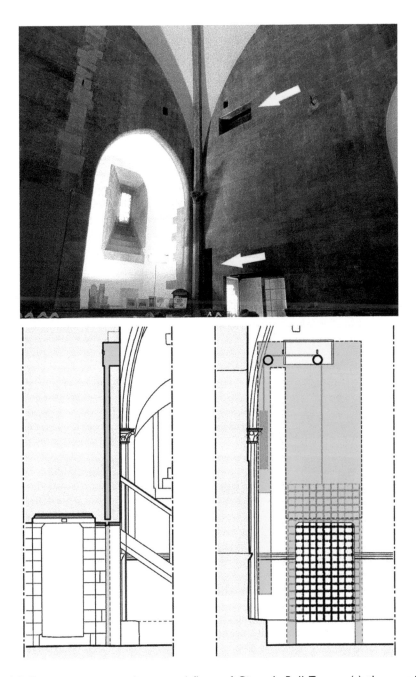

Figure 4.9 Treasure room on the ground floor of Giotto's Bell Tower: (a) the openings in the wall of the current entrance to the bell tower, indicated by arrows, make it possible to inspect the narrow space created for the sliding of a grate, which would have blocked access to the room, and the counterweight compartment; (b) partial section and elevation of the east side of the room highlighting the two openings (yellow) and the slots for the counterweight and the grate (green).

Figure 4.10 On the east façade of the bell tower, the position of the small windows of the toilets created in the wall thickness is indicated (red circles); detail of the windows (below). (right) The overlap of the section and elevation shows that each toilet (blue) is located in correspondence to a room in the bell tower.

The commission Francesco Talenti received in May 1355 to produce a wooden model to correct the chapels and windows seems to have been an initiative that was surpassed by the decisions made in June 1357, which were then included in Talenti's design approved the following August (Guasti, 1887, 81–82, 84). On this date, sudden and resolute guidelines were

issued making the cathedral the focus of attention and actually leading to the forced completion of work on the bell tower with a compromise solution. Under the present-day roof of the
bell tower with its crown of corbels, the start of an additional windowed level can be seen,
which was decided to abandon at this time (Rocchi, 1996a, 139–140).

4.2 Year 1357: the basilican body of Santa Maria del Fiore and the Duomo Nuovo of Siena, two large coeval building phases with different fates

Francesco Talenti took on different but highly important roles at both construction sites. In
Florence, first he was in charge of the construction of the bell tower and then the cathedral
(Guasti, 1887, 94–95, 99). In Siena as a consultant, with Benci di Cione, he was required
to give an opinion on the extremely serious static conditions that had arisen in the basilican
body of the Duomo Nuovo that had just been built (in this regard, refer to the chapter on the
Duomo Nuovo in this same book). Benci di Cione was also already involved in the Opera's
activities in Florence and would continue to be so in the subsequent work to create the pillars
in the cathedral. It is no coincidence that according to a document dated 3 July 1357, Benci di
Cione was asked for an opinion on the foundations of the pillars (Guasti, 1887, 88, 97, 120).

An overview of the chronology of events is also significant in this context. In the spring of
1357, Talenti and Benci di Cione had the opportunity to examine the conditions of the basilican body of the Duomo Nuovo and to assess its vulnerable elements. Essentially, they had an
extraordinary and unique opportunity to reflect not on the basis of a model of the design but
by actually studying a real building of great size which clearly showed the shortcomings and
errors made in the construction choices.

The knowledge acquired by the two master builders at the Sienese construction site must
have been fundamental when starting the construction of the aisles of Santa Maria del Fiore
(Matracchi & Giorgi, 2012, 323–324), also as regards the operational and organizational
aspects of the worksite practices, which as we shall see later were decisive for the stability of
the structure during the execution phases (Giorgi & Matracchi, 2012, 1127–1135). Francesco
Talenti's role as master builder suggests a connection between what happened in 1357 at the
two sites. In July 1357, while the structures of the Duomo Nuovo were being dismantled and
the aisles were being freed from the rubble that had accumulated during the collapses, in
Florence after a long period of little activity at the cathedral construction site, the large bays
were started and completed within a few years, which were much larger than those that had
collapsed in Siena cathedral.

In Santa Maria del Fiore, the change in layout with respect to the sections of the sides of
the cathedral that had already been built is clear (Figure 4.11). The first design was based on
thin external pilasters placed at short intervals, which recall the intention to create bays of
modest depth (Figure 4.12). The slimness of the pilasters, moreover without similar pilasters
on the internal sides of the church, would even suggest the intention to create aisles without
vaulted ceilings (Rocchi, 1988, 69–72). What's more, the positioning of the thin pilasters is
inconsistent with the size of the large bays that would be created.

The portion of the side between the external pilaster and the massive buttress corresponding
to the internal bays of the present-day cathedral, which they decided to create on 13 November
1358 (Guasti, 1887, 120), has a clear resulting width deriving from the dimensions of the bays
of the basilican body established at that time (Figure 4.13). Not surprisingly, the thermographic
imaging carried out on the south side did not detect any traces of an infilled window in this

Figure 4.11 The nave with a view of the crossing.

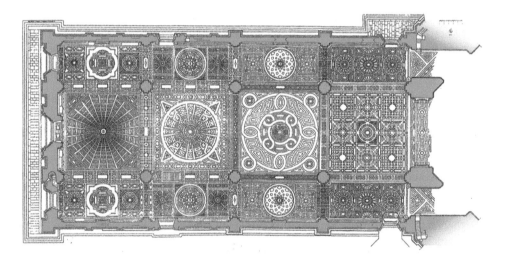

Figure 4.12 Plan of the basilican body.

section (Rocchi, 1988, 70–71). The windows in the remaining bays alternating with the thin external pilasters coexisted with the large interior bays until they were infilled on the inside following a restoration campaign by Gaetano Baccani, the architect of the Opera del Duomo from 1824 to 1860 (Cresti & Zangheri, 1978, 8–9), carried out in 1838–1842. The large fake windows of the first two bays we see today were created on this occasion (Giorgi, 1988c, 45).

Figure 4.13 South side of the aisles, thin pilasters can be seen close to the bell tower, which were then abandoned, and instead, larger buttresses were built corresponding to the bays.

The square bays of the nave of Florence cathedral, delineated by pillar centres placed at a distance of almost 20 m, represent authentic constructive primacy. A significant comparison can be made, to exemplify, between the aisles of the basilican body of Santa Maria del Fiore and those of large churches such as Amiens, Beauvais and Cologne, even using schematic drawings for the latter (Figure 4.14). It is clear that these buildings, which are taller than the Florentine church, have rectangular bays of moderate depth, which significantly reduce the thrusts on the flying buttresses and buttresses. These constructive protections, moreover, do not appear as evident in Santa Maria del Fiore, while they are found in much smaller buildings north of the Alps than those considered.

Returning to the documentation on the Santa Maria del Fiore construction site, the acceleration in the work as of the summer of 1357 should be pointed out; it resulted in the completion of the first two bays with the large vaulted systems in just a few years.

On 19 June, the overall dimensions of the aisles were established, defining the width of the bays.

In July of that same year, excavations of the first two pillars placed near the façade were carried out; in August, the foundations of the next two pillars were laid. At the same time, work started on the ashlars for the base and shaft of the pillars (Guasti, 1887, 98–100, 105, 109). Between November and December 1357, builders laid the foundations of the semi-pillars of the sides corresponding to the first four pillars. On 31 August 1357, Francesco

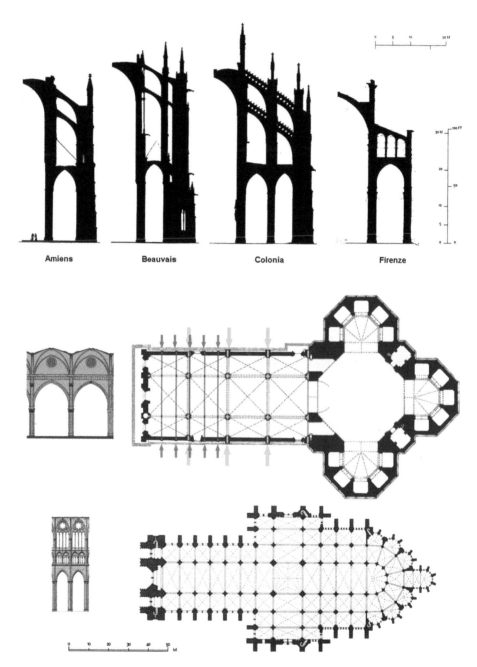

Figure 4.14 (a) Comparison between the structural sections of the cathedrals of Amiens, Beauvais, Cologne and Florence. (b) Comparison between Florence cathedral and Amiens cathedral: in the plan of Florence cathedral, the shallow bays are highlighted (red lines), corresponding to the thin pilasters of the sides (red arrows); but the decision was made to construct roughly square bays in the nave (green arrows).

Talenti undertook to procure the first capital of the columns by January 1358 (Guasti, 1887, 108, 111, 113), but an extension until May was required along with the involvement of his son Simone, who was not paid. In January 1358, they began laying the foundations of the semi-pillars abutting the wall of the façade. In January 1359, the capitals of the pillars and the semi-pillars were paid for (Guasti, 1887, 116, 124).

The foundations of the corner semi-pillars of the façade were partly exposed during the excavations of the ancient Santa Reparata. These semi-pillars were placed outside the basilican body of Santa Reparata, where the floor levels were higher than those inside the church itself. So the top of their concrete casting is close to the current floor levels of the cathedral. The succession of castings with heights of 50/70 cm can be detected in the foundations; moreover, the foundation of the south corner was cast in an excavation that tends to widen as it rises (Marino, 1988, 24–25).

The archival documents show the dimensions of the foundations of the pillars and semi-pillars of the basilican body. For the semi-pillars close to the façade, it was recommended that the foundations be twelve braccia deep, and for the first two pillars, a depth of 10½ / 11 braccia was indicated, reducing to eight braccia for the subsequent pillars, while for the semi-pillars corresponding to the latter, the depths were 9 and 11¼ braccia (Guasti, 1887, 99, 109, 113, 115–116) (Table 4.1). Sometimes the dimensional indications were accompanied by the recommendation to dig "until the good gravel", also specifying to go to "within the water" (Guasti, 1887, 94); it is likely that the knowledge of the subsoil acquired as the excavations progressed led to different depths of the foundations. It cannot be ruled out that the depth of the excavation indicated in the documents also took into account that inside the ancient Santa Reparata, the floor levels were lower than the external ones. This might explain why the foundations of the second pillars are lower than those of the corresponding semi-pillars, which were positioned outside of Santa Reparata.

The problem of the co-existence of the construction site of the new cathedral and the remains of the ancient Santa Reparata emerges in some documentary evidence and in the constructive relationships between the two buildings. The foundation of the first pillar was built close to the side of the south aisle (Figure 4.15). The foundation of the second pillar was instead placed in a more problematic position, close to a pillar of Santa Reparata, which had to be shored up. Attention to the ancient structures of Santa Reparata, which was clearly destined for demolition, was due to the fact that they supported vaulted structures. These covered spaces could be used for construction site activities; in fact, sources mention a chapel "where we work." On several occasions, the documents confirm that foundations were laid "under the vaults," which must have belonged to the earlier church, and in this phase of the works, houses placed up against Santa Reparata were taken down (Guasti, 1887, 99, 101, 106). The foundations of the two subsequent pillars flank the apse of the earlier church and are located in the side chapels.

In this phase, compatibility between the construction works and continuity of worship in the ancient Santa Reparata was certainly problematic and, in any case, necessarily very limited. In December 1358, it was decided to dismantle a part of the roof of Santa Reparata, and the idea of constructing vaults above the remains of the earlier building was considered. The attention paid to continuity of worship in the pre-existing Santa Reparata is evidenced by the frescos found in the remains of the south apse, thought to have been painted in the fifth decade of the 14th century (Morozzi, 1987, 12, 24, 52–53, figs. 17–19); even at a time when the ancient church's fate had been sealed, they did not forgo enriching it with paintings (Figure 4.16). In any case, the delayed start to the construction of the aisles in the new church

Table 4.1 Dimensions of the foundations of the pillars and semi-pillars of the aisles.

Pilaster foundations	Façade wall					First transversal arcade					Second transversal arcade				
	Date	Surface (compensated)		Depth		Date	Surface (compensated)		Depth		Date	Surface (compensated)		Depth	
		br	cm	br	cm		br	cm	br	cm		br	cm	br	cm
North wall	16 Jan. 1357/8	3 × 3	175 × 175	12⅓	720						7 Dec. 1357	6 × 3	350 × 175	11¼	656.5
Central north	16 Jan. 1357/8	5 × 3 / 5½ × 3	292 × 175 / 321 × 175	12 / 12	700 / 700	12 Jul. 1357	(7 × 7)	(408 × 408)	10½	613	31 Aug. 1357	(7 × 7)	(408 × 408)	8	467
Central south	16 Jan. 1357/8	10½ × / 1½	613 × / 87.5	11¾	686	12 Jul. 1357	(7 × 7)	(408 × 408)	11	642	31 Aug. 1357	(7 × 7)	(408 × 408)	8	467
South wall	16 Jan. 1357/8	3 × 2½	175 × 146	12	700							3 × 6	175 × 350	9	525

Figure 4.15 Foundation of the south corner pillar between the side and the façade; the conglomerate was cast on an excavation that reduced in size towards the bottom.

Figure 4.16 Detail of the fresco of the apse of Santa Reparata, executed when work on the new cathedral had been underway for decades.

ensured that the remains of Santa Reparata survived for a long time and resulted in their incorporation into the area delimited by the façade and an initial stretch of the sides. Nonetheless, while on the one hand the remains of Santa Reparata continued to be safeguarded and even enriched, on the other, material from its gradual demolition was used, as attested by the reused ashlars with traces of frescos found in the foundations of the semi-pillars of the counter-façade (Marino, 1988, 25).

   Some main phases in the progress of the works that gradually led to the complete demolition of Santa Reparata can be identified. With the first intervention, which moved back the position of Santa Reparata's façade from the baptistery, the original façade of the ancient church and a section of the aisles were demolished. The subsequent creation of the foundations on arches to thicken the new façade, assuming bays of Santa Reparata of around the same size as the surviving ones, led to the likely demolition of the pillars placed close to this masonry and therefore of the adjacent bay.

The construction of the sides of the cathedral, for the section with pilasters placed on the outside at short intervals, affected the side chapels of Santa Reparata: the side wall was constructed almost in axis with the north chapel, while on the southern side, it overlapped the chapel's external wall. In 1358, marble facing was laid in these parts, which were referred to as "faccie dalato" in the documents.

The subsequent creation of the foundations of the Talentian pillars (1357) had a greater impact on the surviving parts of the ancient cathedral. The foundations of the first pillars were in the aisles of Santa Reparata, whereas the subsequent ones abutted the apse of Santa Reparata taking care not to damage the crypt (Morozzi et al., 1974, 12). Around six decades after the start of the works, great care was still taken to preserve the remains of Santa Reparata, which by this point had become a construction site church, which first survived as a sacellum for the tomb of San Zanobi (until 1439) and later as an ossuary (Nenci, 1996, 18) Coexistence with the remains of Santa Reparata still went on for a long time, given that on 27 February 1375, funds were approved to demolish the walls of the old Church (Caldano, 2018, 84).

On 5 October 1362, it was decided to create two longitudinal arches on the pillars placed on the bell tower side, with the keystones bearing a carved emblem: that of the municipality (a red and white shield) and the other of the judiciary of the Parte Guelfa (an eagle, with a fleur-de-lis over its head, grabbing a dragon with its talons) (Figure 4.17). The construction of these arches was completed in December that same year (Guasti, 1887, LXXXII, 146).

The works continued to progress at a swift pace, and in May 1364, they celebrated the construction of one vault of the aisles. Between September and November of that year, they decided on two important aspects of the nave: the construction of a walkway on corbels,

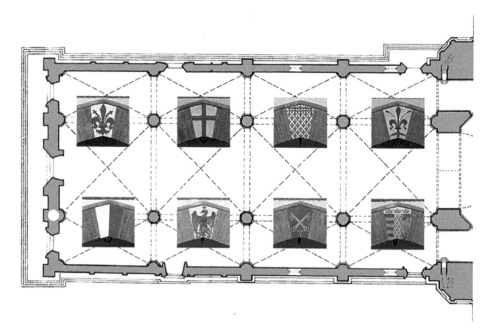

Figure 4.17 The coats of arms at the keystones of the longitudinal arches of the nave.

placed just above the longitudinal arches, and the position of the springer of the larger vaults (Guasti, 1887, 157–160). It is plausible that the vaults of the aisles, in the first two spans, were constructed between 1362 and 1364.

The transition from the nave with rather shallow bays to roughly square bays gave rise to further discussions. Ultimately, it was decided to light the nave with oculi, deciding against the creation of windows. In a resolution dated 20 December 1364, in addition to reiterating the choice of oculi, it was specified that each bay would be illuminated by one window and not three, alluding to the three openings existing in the external wall of the aisles, corresponding to the first bay (Guasti, 1887, doc. 126).

The inevitable indecision during the phase of shifting to the solution of large bays also seem to be confirmed in the construction of the first two longitudinal arches. In the garrets of the aisles, the walls above these arches actually have pilasters that reflect the position of the thin ones on the external walls (Rocchi, 1988, 69). In this phase, they did not rule out perhaps continuing them up to the external side of the windowed walls of the nave; not surprisingly, the final choice to abandon three windows and instead create just one oculus was made thereafter, towards the end of 1364.

The building of the two large vaults of the nave was also concluded with resolve and the necessary caution. In July 1365, after having completed the first groin vault, an opinion was sought on the centring and the walls below the second, the completion of which took place in March and was celebrated in May 1366 (Guasti, 1887, docs. 131–136).

The following July, an important resolution was made: the consultants of the Opera established that the works would continue with four bays and that in the sides of the aisles a window would be created for each bay delimited by massive pillars. In fact, they decided not to continue with the design started in the sides of the first two bays, in that the windows positioned between the thin pilasters "don't look good" (Guasti, 1887, doc. 141). This final decision marked the surpassing of the design that had been followed before creating the nave with large bays in the summer of 1357.

4.3 The façade: architecture, structure, functions and alterations

The changes linked to the new basilican body left significant masonry evidence also in the spiral staircase of the façade, mentioned in a document dated January 1358 concerning the laying of some foundations (Guasti, 1887, 115), abutted by a semi-pillar of the nave (Figure 4.18).

The wall structure of the spiral staircase is characterized by pietraforte ashlars of a regular height equivalent to the rise of the steps of nave (Figure 4.19). Starting from the bottom, the first window is at a height of +4.59 m from the church floor, infilled on the outside due to the construction of the Talentian semi-pillar, and therefore incompatible with it (Figures 4.19B, 4.20B) (Giorgi & Matracchi, 2006b, 283–290).

The next window in the spiral staircase (at +10.18 m) is completely infilled by a stone and brick infill (Figures 4.19C, 4.20C). A test checked the conformation of this window which, unlike the previous one, might have co-existed with the Talentian semi-pillar and was likely infilled due to the crack that appeared in its lintel.

The orientation and conformation differences of the two windows examined highlight an important building phase; the Talentian semi-pillar must have been built when the spiral staircase had reached a height of between the first and second window. It could be assumed

Figure 4.18 Counter-façade of the nave.

that starting with this latter, the spiral staircase was constructed considering the Talentian semi-pillar. At the same time as the construction of the Talentian semi-pillar, at the foot of the spiral staircase, a new entrance door was also created, but repositioned with respect to the original one (Figures 4.19A, 4.20A).

Another aspect concerns the constructive relationship between the Talentian semi-pillars and the loggia next to the lunette of the counter-façade portal. On each side of the portal is a blind loggia with three arches; the outer ones are half infilled by the semi-pillars.

Taking into account this unique feature and the state of conservation of the stone facing inside the façade, characterized by a significant alteration that can be attributed to long exposure to bad weather, we can speculate that when the semi-pillars were built, the masonry of the façade had reached the level of the blind loggias, with the one whose arch is adjacent to the spiral staircase being left incomplete while waiting for the builders to continue constructing the staircase. On the other hand, as the stone arches of the loggia were set into the masonry, the arch adjacent to the spiral staircase could have been partially completed, without the springer on one side.

In the spiral staircase, if we take into account the window in the lower position and assume a pilaster compatible with it, it can be confirmed that the outer arches of the loggia, now interrupted, could have been completed in full (Figure 4.21). In other words, a loggia was constructed that on the outer side, should have abutted thinner pilasters than the Talentian semi-pillars actually there, or, more likely, the masonry above the longitudinal arches set on shorter pilasters than those there today. The connection between the dosserets of the capitals of the Talentian pillars and the archivolts of the arches of the blind loggia also confirms that the loggia already existed: the dosserets were chiselled and plastered to make them adhere to the protruding bands of the archivolt, which continue intact behind these dosserets.

But other variations became necessary during the construction of the Talentian semi-pillar. In October 1358, discussions were held on whether to place the capitals of the pillars at a higher level (Guasti, 1887, 120), at the time planned to be seventeen braccia from the floor level (9.86 m), but then later raised to the current height of 11.35 m (19.5 braccia).

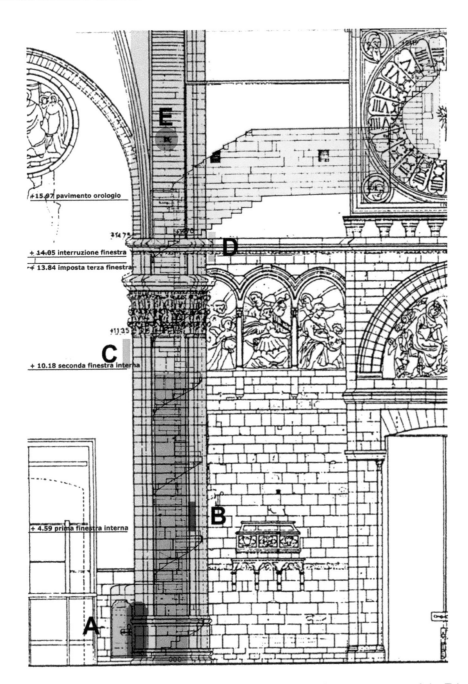

Figure 4.19 First stretch of the spiral staircase of the façade; the construction of the Talentian semi-pillar involved the closure of the original door (A, red) and a window
(B); the second of the two windows (C and D) compatible with the semi-pillar
was not completed in that it was decided to shift the staircase towards the
centre of the façade; this would provide better anchoring for the longitudinal
tie (E).

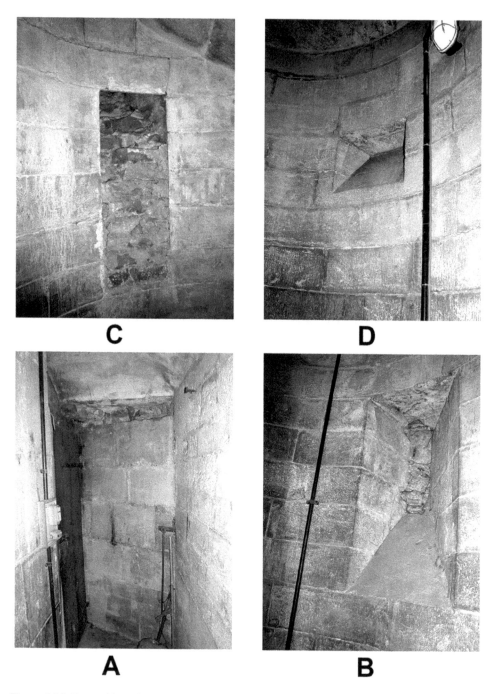

Figure 4.20 Door (a) and windows (b–d) of the first stretch of the spiral staircase of the façade.

Figure 4.21 The full extension of the small arch to the left (red) was achieved with a narrower original pilaster (grey).

Before deciding on this change, the construction of a third window in the spiral staircase (at +13.84 m) was started, but only the bottom part of it was completed as it was then interrupted by the perimeter wall of the spiral staircase (Figures 4.19D, 4.20D).

At the same time as the semi-pillar was built, in order not to weaken the area supporting the longitudinal arches of the aisles where the longitudinal tie should have also been inserted (Figure 4.19E), it was also decided to move the spiral staircase towards the centre of the façade by means of a corridor placed partly on the internal overhang towards the nave created by the loggia with prominent archivolts.

The walls on the sides of the corridors were of different thicknesses: 86 cm towards the inside of the church, while on the outside, the original thickness was around 25 cm of rustic masonry, as the wall had not yet been cladded. The builders were certainly not ignorant of the fact that the central part of the façade with the portal and the oculus (or large window) above, in all likelihood planned, was the weakest structural point where, in large and small buildings, the tension caused by thrusts or subsidence is almost always discharged resulting in the formation of vertical cracks. Adding the void created by the spiral staircase between the two openings would have made this area of the façade even more vulnerable.

Some constructive solutions were therefore adopted. The second spiral staircase, with respect to the first, was positioned closer to the inside of the wall. Both staircases have the same diameter; however in the second, placed in a position of the wall that was less rigid due to a lack of semi-pillars and buttresses, the central column was enlarged (from 20 to 27 cm), thereby reducing the width of the steps. Moreover, the area surrounding the second spiral staircase was strengthened by a wall thicker than the initial part of the corridor: the external infilled door positioned at the end of the corridor and the first window of the spiral staircase enable us to identify the Medieval external limit of the wall, excluding the 19th-century completion of the façade (Rocchi, 2006a, 249). Two areas of the façade were thickened, one on the outside limited to the width of the portal and the other inside extended to the entire nave, and, again on this side, a further wall addition concerned just the portal.

A comparison between the drawings by Poccetti (Figure 4.22) and De Fabris (Figure 4.23) and the building phases of the façade identified in the first stretch of the spiral staircase shows that the first window of the spiral staircase (at +4.59 m), not compatible with the Talentian semi-pillar, corresponds to the first register; the next window (at +10.18 m) congruent with the Talentian semi-pillar, is halfway between the two subsequent registers. Moving from the level of the second window to the interrupted one (at +13.84 m), between the spiral staircase and the central portal, a room surmounted by a segmental arch was created following the drawing by De Fabris. In the same recess, Poccetti placed two niches with archivolts, divided by fluted pilasters, while in the painting, The *Plague of Florence in 1631*, rustic niches can be seen.

When the decision was made to interrupt the spiral staircase and to move it towards the centre of the façade, the level reached was not yet high enough to get over the external niche surmounted by the segmental arch of the façade. So the change of position was made by adding an initial straight ramp that led into the corridor, which in turn had a slight slope up to reach the level where the second part of the spiral staircase started. It should not be underestimated that this raising of the corridor left more space for the central portal.

The way of reconciling the construction and compositional requirements, as well as those linked to the accessibility of the different areas, also necessary for subsequent maintenance, seems to indicate that Francesco Talenti played a leading role in all of this, finding himself with a construction site situation that was, in some aspects, similar to that of the Bell Tower of Santa Maria del Fiore where, by alternating spiral staircases with straight ramps, he managed to create large windows – monofore, bifore and trifore – on all sides, while at the lower levels created by Giotto and Andrea Pisano, the openings were smaller and sometimes overlapped with stairs (Rocchi, 1996a, 137–138).

The parietal structure of the cladding of the façade drawn by Poccetti would seem to reflect the observations made earlier concerning the phases of the initial spiral section. The part of the first register of the façade, with large niches and solid buttresses, seems linked

Figure 4.22 The medieval façade designed by Gaetano Poccetti before its demolition in 1587.

to the phase prior to the Talentian semi-pillar. The two subsequent registers of the façade, joined to buttresses left empty for statues, could both (the second necessarily was) have been executed after the window of the spiral staircase was adapted to the Talentian semi-pillar: it is plausible that as the works progressed the execution of the masonry structure, and the marble cladding was not contemporary, as attested by many examples in which the laying of the incrustations was postponed. Moreover, above the spiral staircase, Poccetti's drawing has a further level of buttresses delineated by columns with a statue inside. The choice not to use solid buttresses to create niches was reasonably linked to the decision to move the spiral staircase and therefore to place the longitudinal arches on solid façade masonry, which was more reliable also for anchoring the ties of the same arches.

The sequence of phases linked to the relationship between the stairs and the Talentian semi-pillar and the subsequent change of position of the staircase to the central part of the

Figure 4.23 (a) The survey made by Emilio De Fabris before the construction of the 19th-century façade. (b) In the rustic façade designed by De Fabris, the colours highlight the change in thickness of the wall, from the thickest part at the portal (purple) to the thinner parts of the garrets of the aisles (light blue) and the nave (yellow).

façade is also indirectly confirmed in documents: it was not until 1359 that Alberto Arnoldi was entrusted with the execution of the arch of the central portal and, not surprisingly, in the assignment of this task, the builder was invited to carry out the work in close collaboration with Francesco Talenti (Pomarici, 2004, 60).

After the spiral staircase section in the centre of the façade, access to the garrets of the nave was obtained by a straight ramp moved to the inner side of the façade; this way, the smaller width of the staircase at the oculus level meant the wall thickness could be reduced, which is clearly visible in the drawing by De Fabris. This was also achieved by the choice to place the subsequent straight ramp, which was introduced into the garret of the nave, directly on the extrados of the first groin vault of the nave.

The rustic façade represented by Poccetti apparently had three masonry planes: the innermost of the two large twin recesses, which seems to correspond to a narrow niche located further to the left and also present on the right in De Fabris's drawing: the one corresponding to the oculi of the aisles and nave and the outermost of the foreparts above the portals. In De Fabris's drawing, the level of foreparts coincides with that of the side oculi and the masonry above them, while the dimensions of the recesses in the masonry are highlighted here by the technique of representing shadows. The blind arches are set back by around 90 cm, coinciding with the measurement stated by the architect for the depth of these elements; the narrow niches alongside the smaller portals are 20–25 cm deep, and the depth of the horizontal offset seems to be smaller still, just below the oculus of the nave. Several sources indicate that the greatest wall thickness before the 19th-century interventions was around 2 m (Giorgi, 1988b, 33). The wall was a total of 3.5 braccia, with large niches 1.5 braccia deep.

Beyond the differences between the two drawings, the most setback masonry plane seems to have been created to position statuary inside niches: the two large arches are a continuation of the coupled niches of the first three registers next to the central portal; the narrow niche is located over the niches obtained between side portals and buttresses. In other words, the façade represented a game of indentations (the innermost walls) and protrusions (buttresses with solids and voids) where the statuary would have been placed. It should be pointed out that the apparently unique solution to use a pillar to divide the large niches from the much narrower side ones was entirely justified by constructive reasoning, which advised against the creation of voids in this part too, where the longitudinal arches rest and the anchor bars of the ties were fixed.

The 14th-century rustic wall facing, which can still be detected today, is identified as seen at the level of the clock corridor: in particular, the position of the back wall of the blind arches and the external row of the forepart have been identified. The external facing of the 14th century façade can also be identified at the level of the garrets of the nave and of the aisles, where it is possible to measure the first wall through the putlog holes. From the metric data obtained in the garrets of the nave, the thickness of the 14th-century wall, measured considering the discontinuities of the wall structure also visible inside the putlog holes in the counter-façade, was around 80 cm and below them was an external offset of 45 cm according to the measurements of Del Moro (Giorgi, 1988a, 40), also attested by 19th-century photos prior to the construction of the new façade. The end walls of the garrets of the aisles, here also measured in the putlog holes, had a rustic thickness of around 43 cm. It should be considered that the garret walls were reinforced by pillars.

In the first two windows of the second spiral staircase, a discontinuity similar to that observed in the putlog holes of the garrets can be seen, allowing us to identify the total thickness of the Medieval façade at this point – the thickest of the whole façade due to the presence of areas where the wall was thickened inside and out – as 244 cm, just over four braccia, versus the two braccia measured in the side walls. The overall thickness, not considering the 40 cm overhang of the counter-façade wall of the nave above the blind loggia, is equivalent to the 2 m indicated by all ancient sources, including De Fabris (Giorgi & Marino, 1988, 51) (Figure 4.24).

The architectural diagnostics of the façade have highlighted an extremely complex Medieval architectural history, characterized by variations and changes of mind and a great variety of masonry structures with some parts made thicker and other parts made thinner. This allowed the builders to create spaces for a complex system of niches for statuary, only partially completed as documented by Poccetti, and to guarantee the stability of the structures creating precise constructive relationships with the aisles. In addition, there was a complex system of stairs, which led from the church to the garrets of the nave, based on spirals and straight ramps, intimately linked to the variations in the wall thicknesses.

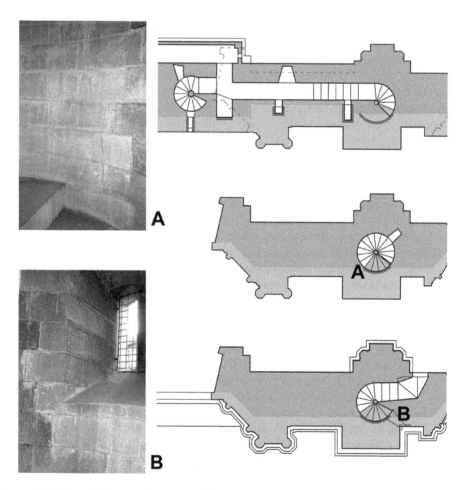

Figure 4.24 Plan of the spiral staircase from the level of the church to the height of the clock on the counter-façade; the 19th-century walls and the restoration of the external part of the spiral staircase are indicated (pink).

The complex genesis of the Medieval façade, although the oculus in the façade was worked on until 1404 and 1412 (Guasti, 1887, docs. 424, 462), was followed by 16th-century modifications, which dismantled the initial Medieval marble façade without succeeding in the intention to execute the design by Bernardo Buontalenti, winner of the competition endorsed by the Grand Duke Francesco I (Bevilacqua, 2015). They settled for the 17th-century façade painted on a brick wall, until in the last quarter of the 19th century, it was also dismantled to create the current façade designed by Emilio De Fabris (Rocchi, 2006b, 374–386). The impact of this intervention on the Medieval structures was considerable (Figures 4.25 and 4.26). A large portion of the first section of the spiral staircase, with the adjacent wall, was dismantled and reconstructed to replace the steps cut in the 17th century and to reintegrate

Figure 4.25 The 19th-century rustic masonry of the façade with the route of the medieval spiral staircase (green) up to the final stretch (purple), which rests on the vault of the nave, is highlighted; a section of the stair was demolished (yellow) and reconstructed to the side (red) in order to create the façade niche (light blue) that houses the statue of the Madonna and Child.

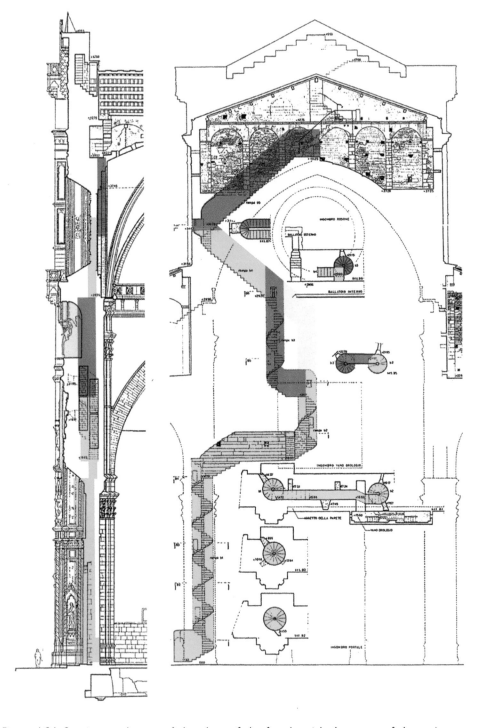

Figure 4.26 Sections and parts of the plans of the façade with the route of the stair.

the external facing. A section of the spiral staircase, between the clock and the balcony of the nave, in order to leave space for a niche with a statue of the Madonna and Child by Tito Sarrocchi, was even deviated and rebuilt to the side by breaking through the wall. If we consider the stretch of the dismantled staircase, documented in De Fabris's drawings, the construction of the niche for the statue of the Virgin and the new stretch of staircase, it is clear that large sections of the Medieval masonry were demolished to create the new structure with the niche alongside the spiral staircase. This intervention significantly reduced the wall mass that Medieval builders had wanted to be thicker, creating internal and external wall thickenings. Even the 19th-century phase significantly increased the complex and heterogeneous character of the façade wall. Work was carried out on the structures at different depths to create a new rustic external wall with overhanging rows extending for the entire front, as shown in a drawing by De Fabris; marble incrustations were then added, as seen on the present-day façade.

4.4 The construction of the aisles: the stability of the structures during and after construction

A few months after having examined the critical conditions of the basilican body of the Duomo Nuovo in Siena, Francesco Talenti became head of construction of the aisles of Santa Maria del Fiore, which were even larger than those of the Sienese church. Considering the centre-to-centre distances of the pillars, the bays of the nave in the Duomo Nuovo are around 15.60 m wide and 12 m deep (with some depth variations); the height of the groin vaults where they join the façade wall is 33.90 m. In Florence cathedral, the bays of the nave are around 19.40 m wide and 19.85 m deep; the groin vaults where they touch the walls reach a height of around 39.55 m, and at the extrados of the keystone, they rise to 42.50 m (Figure 4.27).

For Francesco Talenti and Benci di Cione, who both produced a survey report on the Duomo Nuovo while engaged in works promoted by the Opera di Santa Maria del Fiore, the Sienese experience must have been influential in the pivotal year of 1357 which marked the decisive start of work on the aisles of Florence cathedral, which had dwindled until then.

All the constructive shortcomings observed and complained about by the two experts, referred to in a partial but very interesting testimony in Benci di Cione's report, namely the fact that ties were not used, the poor masonry quality of the pillars, asymmetry in the walls, and a lack of care in executing the masonry connections (in this regard see the Section 3.3), had to be carefully considered when developing the appropriate construction solutions to be used for the aisles of Santa Maria del Fiore in order to create a design that was structurally even more ambitious than the Sienese one (Figure 4.28).

The attention paid to the construction aspects is apparent in several documents. In October 1358, there were discussions over whether the courses of the semi-pillars should be continuous with those of the masonry and if the walls should be made of ashlars rather than stones. It was recommended that the springer of the larger vaults be placed below the level of the corbels of the walkway of the nave (Guasti, 1887, 118–119, doc. 120). When it was time to construct the second vault of the larger aisle, above it, the master builders advised reducing the thickness and limiting the height of the masonry (Guasti, 1887, doc. 131). This recommendation was fully implemented as the wall was made thinner and at the same time stiffened by pillars; moreover, its height allows the trusses to sit a bit higher than the extrados of the vaults. For the larger vault, the creation of a strong connection between the corbels, the vault itself and the masonry was also suggested.

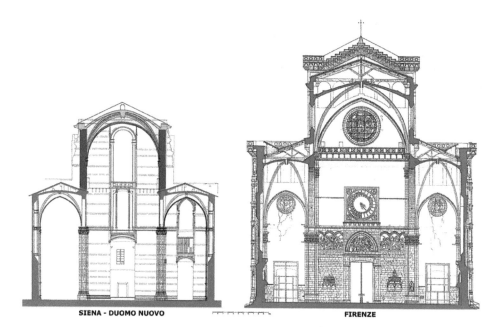

SIENA - DUOMO NUOVO   FIRENZE

Figure 4.27 Comparison between the cross sections of the Duomo Nuovo (a), with the hypothetical completion of the vaults of the nave and the right-hand aisle, and the cathedral of Santa Maria del Fiore (b).

The builders had to tackle the complex static problems posed by the large dimensions of the three vault-covered aisles. In addition, they paid particular attention to the temporary incompleteness of the structural framework encountered during construction. As the works gradually progressed, particular instability conditions were created due to stress on structures that were very different to the final ones.

While building the first two bays, where the effectiveness of the solutions developed was put to the test, three main and different static conditions had to be addressed. In the first phase of works, the aisles were constructed and covered by vaults, as if they were two separate building parts connected by the façade wall at one end. Construction continued with the transverse arches and vaults of the nave. Note, however, that the final constructive structure could only be achieved after the entire basilican body with four bays had been built. It should also be added that the maximum efficiency of the vertical loads, which shifted the thrust vector downwards rather than sideways, was only obtained after the trusses of the nave were put in place, some of which have been dated to the mid-15th century (Bebber, 1988, 61–62). Moreover, the function performed by the ties of the trusses, namely that of providing contrast and friction, in stabilizing the top part of the flanks of the nave should not be overlooked.

All the construction difficulties linked to the size of the building and to the particular stability conditions encountered in the different work phases were overcome by setting up temporary devices. In particular, metal and wooden ties were used, which in some cases were removed and in others conserved, and actually became inoperative as the completion of the church progressed. None of this, however, prevented unexpected difficulties occurring at the

Figure 4.28 Florence Cathedral, view of the vaults of the north-side aisle in relation to those of the nave.

Santa Maria del Fiore construction site too. In order to better illustrate the difficulties faced by the builders, it is worth going over the key steps of the creation of the basilican body.

The aisles were equipped with ties in the longitudinal and transversal arches, with a rectangular section of 7×5 cm (Giorgi & Marino, 1988, 53). These ties foreshadowed the execution of the vaults and were placed at the same level so they could be anchored to the same anchor bar inserted into the masonry above the pillars. The placement of the ties at the same level resulted in a unique constructive feature: the longitudinal ties were placed, as usual, at around 1/3 of the rise of the large arches, while in the transverse arches, which had a much smaller span, the ties were consequently placed at well over half of the rise. But in this case, to avoid placing the tie in a position that was too unusual, stilted arches were created with a real springer much higher than the apparent springer on the dosseret. Great importance was clearly given to the coplanarity of the ties in order to create a more rigid plane to counteract the tensile stresses generated by the thrusts of the arches (Figure 4.29).

The transverse arches were also shaped like diaphragm arches, the structural function of which was to create a connection between the flanks of the aisles and the masonry above the pillars and, at the same time, to act as a masonry strut interposed between these walls.

In this execution phase, in the central part of the longitudinal arches, the wall sections were small and therefore particularly subject to deformation in that the level of the walkway on corbels had not yet been reached. In the arch keystones, about 25.84 m high and with a span of almost 20 m, there was a masonry section 1.10 m wide and roughly 1.24 m

Figure 4.29 Part of the longitudinal and cross sections of the north-side aisle with the visible
tie system highlighted (red).

high. But we should not disregard how these walls were strengthened by the pilasters in
the most lateral areas, incorporated into the garrets of the aisles (Figure 4.30).

The proximity of the putlog holes to the extrados of the vaults can be observed in the garrets of the aisles. This confirms that above the longitudinal arches, masonry was constructed
up to the support level of the corbels of the nave, using scaffolding that rose from the ground.

The height of the keystones of the transversal arches was much lower than that of the
longitudinal arches. The longitudinal section of the aisle shows that the particular position
of the diaphragm arches was also linked to the decision not to use the filling material of the
spandrels; in fact, at the extrados, almost the entire longitudinal development of the brick
vaults can be seen.

The apex of the transversal diaphragm arches was dictated by the choice to rest the pillars
of flying buttresses there in order to counteract the vaults of the nave (Figure 4.31). This meant
that some of the thrust was channelled to a rather low level on the flanks. So when the walls
above the longitudinal arches were built, in addition to stiffening pilasters, the pillars of the
flying buttresses abutting these walls were also constructed at the same time. In this regard,
some parts of the façade corresponding to the garrets of the aisles are significant, where the
wall thickness was reduced with a marked wall offset to leave space for the pillars of the flying
buttresses, which were infilled on the outside by a thin wall of around 40 cm. It is clear that in
this case, the cogent problem of creating a lightweight structure above the diaphragm arches
did not arise; here there is a continuous underlying wall with an oculus and a side doorway. In
any case, they did not forgo making the masonry lighter by creating an infilled flying buttress,

Figure 4.30 Longitudinal section of the north aisle. The image highlights: the diaphragm arches (red), flying buttresses (green), the pilasters of the side of the nave (light blue), the stone tie (blue) and the corbels (orange). The yellow area indicates the first two bays that were built, the temporary completion of which is clearly visible in the garret (red arrow).

but it was not as wide as the isolated ones. The walls were lightened in a similar way also in the part of the façade corresponding to the nave where there are pillars that support internal arches accompanying the entire perimeter of this level, except for the East side, where the massive pillars of the crossing were later built.

Small through holes were created at the base of the pillars of the buttress of the first archway of both aisles, probably for the placement of ties which the builders then decided not to insert and then did not create the slot at the base of the pillars of the subsequent flying buttresses. In any case, the appropriateness of giving the transverse diaphragm arches an additional tie positioned in the garret, parallel to the visible one of the soffit, was initially assessed.

The particular height of the transverse arches of the aisles, linked to the construction of the flying buttresses above concealed in the garrets, is also evidenced by a comparison with the aisles of Santa Croce church in Florence, which for the most part had already been constructed by the 1360s (Carbonai et al., 2004). The width of the nave of Santa Croce, covered by wooden trusses, is comparable to that of Santa Maria del Fiore. But there, the transverse arches of the aisles are the same height as the longitudinal ones (Figure 4.32). It is also significant to recall the surviving aisle of the Duomo Nuovo of Siena, where the longitudinal and transverse arches are of equivalent height. With the solution developed for Florence cathedral, between the longitudinal arches and the flanks, a transverse connection was created at an intermediate level between the capitals of the pillars and the gallery, in a much lower position than the longitudinal arches; flying buttresses were also added to counteract the thrusts of the vaults of the nave.

Figure 4.31 The cross section shows the construction works up to the flying buttresses positioned on the diaphragm arches of the aisles.

The vaults of the aisles were made of brick and in the wider part were one brick thick, becoming one-and-a-half bricks thick only in the lower parts, which are visible close to the diaphragm arches (Figure 4.33). To construct them, transverse iron ties were set up, partially visible from the garrets and positioned close to the transversal axis of the vault. The ties have rather thin sections (2.5 × 2.5cm) and are anchored, on the internal side, in the area between

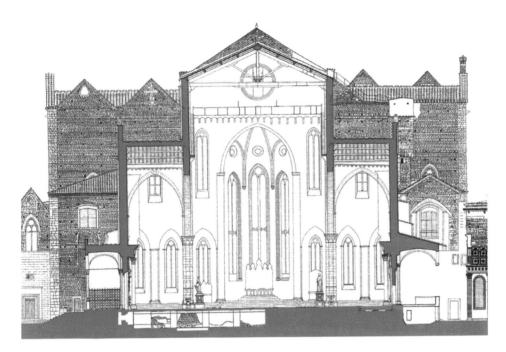

Figure 4.32 Cross section of the basilica of Santa Croce in Florence.

the top of the large coats of arms at the keystones of the longitudinal arches and the cornice at the base of the corbels above, and on the external side around the springer of the corbels incorporated into the garret. In the most central part of the webs, these ties were incorporated into the brick masonry of the vaults built on centring. These ties consist of two distinct inclined bars, probably connected by eyelets and dowels at the height of the keystones.

The visible sections of the two tie segments vary due to different circumstances: the different curvatures of the transverse directrices of the webs, those on the outside are more rampant; moreover, on the extrados of the internal webs, close to the aisle wall, a walkway was created with rustic flooring made of brick and abundant mortar, which at times skirts the ties and in other cases incorporates them. As a result, stretches of the ties are more visible, mainly in the outer webs (Figure 4.34). The different curvature of the transverse webs gives the groin vaults a slightly rampant shape crosswise, which the builders must have considered an effective aid in offsetting the thrusts from the nave.

Special investigations with a metal detector revealed that the visible metal bars continue inside the vaults. The presence of a bar fully embedded into the external side of the first south bay was also detected.

These ties acted as temporary works: the lack of additional vertical loads created the conditions, above all in the keystone area of the large-span longitudinal arches, for transverse deformation due to the thrust of the vaults of the aisles built following a curved directrix.

The corbelled gallery pertaining to the first two bays of the nave was executed in 1364, also with the intention of strengthening the structures (Guasti, 1887, doc. 119). This function can be explained by taking into account that at the level of the cornice supporting the corbels, a row of squared ashlars was created, which can be seen on the side of the garret. These were

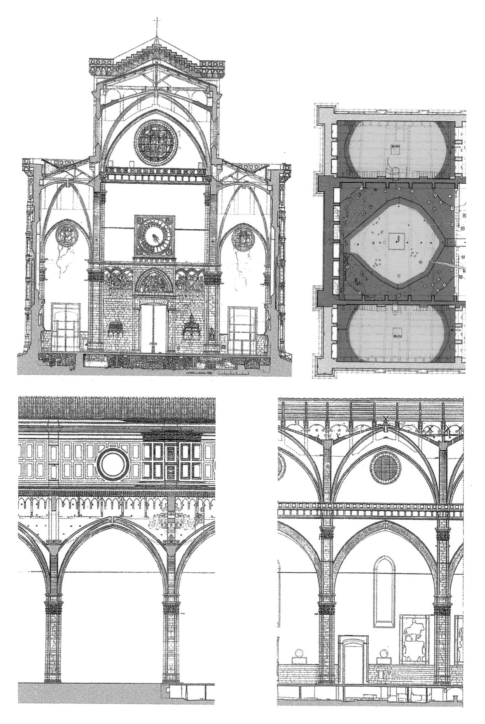

Figure 4.33 The sections and the plan show the thicknesses of the vaults one-and-a-half-brick thick (red) and those one-brick thick (green).

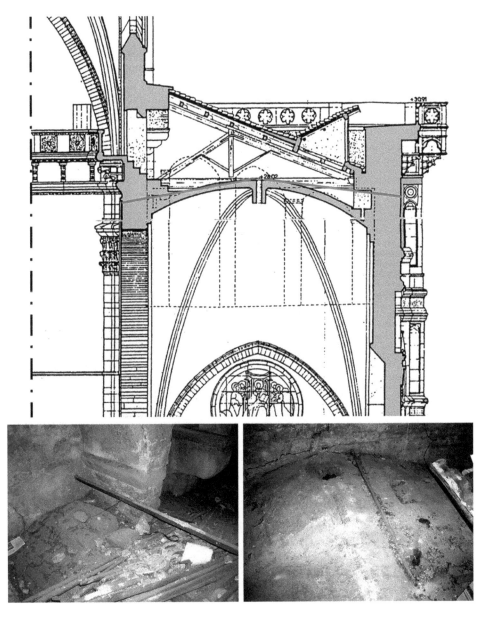

Figure 4.34 (a) The position of the ties partially inserted into the vaults of the aisles is indicated, with the anchor bar corresponding to the keystone of the longitudinal arch. Connection of the ties with the outer side (b) and the inner side (c).

sandstone ashlars, in all likelihood, connected to each other by cramp irons placed on the upper side and embedded into the masonry. This course acts as a longitudinal stone tie and creates a band in the masonry which helped to improve the distribution of the vertical loads that would have been produced as the building works progressed. The importance of this stone tie is evidenced by the fact that with cramp irons of considerable length placed on the side of the garret, some fractured stone ties were reinforced.

The Talentian corbelled gallery and the corbels on the two sides of the garrets were created in continuity with the vaults of the aisles, and the flying buttresses with three archways and the pilasters of the garrets above the longitudinal arches were completed. The close execution relationship between these elements and the adaptations that this entailed are evident. The pilasters of the garrets, corresponding to the position of the external ones of the flanks, assumed a width (146 cm) equal to the span of one arch and two corbels of the gallery above, therefore greater than the external pilasters (116 cm). The span of the arches of this gallery, in turn, was contingent upon the presence of the flying buttresses placed on the diaphragm arches of the aisles. The flying buttresses at the end of the first bay abut one side of the garret pilaster; the wall pillar of the flying buttresses of the second bay harmonizes with the span of the arches, while the garret pilasters had been abandoned by this point (Figure 4.35).

Figure 4.35 (c) Garret of the aisles; flying buttress with the side abutting the external flank. (a) Corbels of the outer side. (b) Flying buttress abutting the inner wall (green), with the adjacent pilaster (light blue) and corbels (orange).

Concordance was also created between some wall thicknesses: the thickness of the garret pilasters was equivalent to the overhang of the first of the three quarter-circle brackets of the aforementioned corbels. The garret pilaster only varied in size where it connected with the façade: at this point, it was not as thick, and its width extended for about one and a half arcs of the corbels.

The corbels of the garret of the aisles, on the inner side, are formed of three brackets and have an overhang of about 90 cm, while on the side of the external flanks, there are two brackets, and the overhang is 70 cm. Both act as maintenance walkway and are still used today. The inner corbels, however, given that they abut the wall of the nave, had to be wider. Only on the outer side was it possible to follow the thickness of the underlying walls of the flanks. In any case, in this last part, the walkway was not accessible before the projecting gallery resting on corbels was built in the 16th century. In fact, the external corbel crowning was only created from the start on the apsidal galleries (Giorgi & Marino, 1987, 63–68).

The walls that rise above the longitudinal arches, which are only around 90 cm thick, were considerably stiffened by the garret pilasters, the stone tie, the gallery of the maintenance walkway and the flying buttresses.

Nevertheless, while completing the flying buttresses, an additional temporary work was set up, linked to the static conditions encountered as the work progressed; in this phase, they had arrived at the construction of the vaults of the first two bays of the aisles and the masonry of the nave rose to just above the springer of the vaults of the nave.

In similar conditions, the three sloping arches of the flying buttresses were given tie rods, which traced their inclination. The use of tie rods in this case should be considered unique, taking into account that the flying buttresses serve to counteract the thrusts from the vaults of the nave, which would have rendered them ineffective. But, in this phase of the work, the absence of vaults in the nave and the lack of the additional load of the walls of the clerestory led to the fear that longitudinal wall supported by pillars would bulge, with possible movements towards the nave. So, after having placed transversal ties inserted in the centre of the vaults of the nave, it was deemed appropriate to place others in correspondence to the flying buttresses in order to add further connections, with a temporary function, between the walls flanking the aisles. For these anchors, placed almost at the same height, ties with a rather thin square-section were used (with sides of 3.5 cm, a comparable size to those inserted into the vaults of the aisles) as evidence of the fact that they hoped to counteract the minor inward thrusts.

This led to the construction of the large vaults of the nave, which represented the most complex phase of the works. Flying buttresses had been planned, but they could not rely on the load contribution of the wall masses of the flanks of the nave at garret level and of the roofs, which guide the thrust (vector) downwards rather than sideways, as these parts were still to be built.

On 18 July 1365, Giovanni di Lapo Ghini was asked to provide an opinion on the centring and on the vault constructed in the first bay of the nave (Guasti, 1887, doc. 131). In the same resolution, positive conclusions were drawn regarding the centring and the general management of the works, but there were also recommendations on the construction solutions to be implemented.

One recommendation concerned the close constructive relationship between the corbel of the vaults, the vault itself and the masonry. It is likely that the springer area of the vault was constructed with horizontal rows directly connected to those of the surrounding masonry. As mentioned earlier, one of the shortcomings pointed out by Benci di Cione regarding the Duomo Nuovo in Siena was precisely the insufficient connections between different masonry elements.

A further recommendation suggested the insertion of a tie placed five braccia below the apex "of the archivolt of the vault", a height that would roughly correspond to the traces of a tie found in the transverse arch between the first and second bays. This tie was placed very high up with respect to the arch springer, whose function might have been temporary. Temporary ties can be seen, at a height just above the gallery, in all the transverse arches. The fact that a trace of a tie can only be found in the first transverse arch, at a very high level of the intrados, is evidence of the uncertainties that arose during the execution of the first large groin vault.

Evidence of temporary ties on the intrados of the transversal arches is found in the protruding pieces of metal – with sections of 5 × 7 cm, 5.5 × 7.5 cm and 5.5 × 7 cm – the probable remains of hooks or eyelets, and in an eyelet almost completely embedded into the masonry. These hooks, when placed at the edge of the intrados, are next to a patch of mortar indicating the slot for a transverse dowel to lock the flanking eyelets into place. Above the hooks, there are traces of additional metal anchoring with a smaller section than those below.

All of this indicates that the nave was equipped with temporary ties, which were probably wooden with metal anchors with eyelets at the end (Figure 4.36). The higher anchors must have been hooks put in place to lift the beams by means of ropes, in order to allow the connection with the hook, or with the eyelet by means of a dowel, set up on the intrados.

In the nave, the original plan was not to have permanently visible ties but only temporary ones set up to address the insufficient vertical loads, due to the incompleteness of the walls of the flanks and the lack of roofing.

Above the temporary ties, at the level of the garrets of the nave, there were tie rods made up of three pieces crossing the top of the transverse arches. The two lateral sections are entirely visible, while the central one is almost completely inserted into the arches, except for a short section that ends with eyelets (Figure 4.37).

These reinforcements were placed much higher than the springer of the groin vaults, and at this stage of the construction, their function must have been mainly to block possible movements of the masonry of the sides of the nave until they could rely on the stabilizing function of the large roof trusses.

It was only with the completion of the vaults of the first two bays of the nave in 1366 that the thrusts towards the aisles began to act, activating the countering function of all the flying buttresses. But the loads were still partial as the vaults of the first two bays were still resting on centring. These vaults were built with the same thicknesses as those of the aisles; but in the larger vaults, the part that was one-and-a-half bricks thick was considerably extended, consequently reducing the central part to one brick thick (Figure 4.33).

The wall trace that indicates the end of the first two bays is clearly visible in the garret of the north aisle, beyond the flying buttress, where a wall disconnection with an irregular sub-vertical trend can be seen. This short stretch of interrupted wall rose from a small part of the third arch and projected beyond the second pillar. The stability of this corbelled overhanging part of the wall also had to be ensured by inserting metal bars into the mortar beds of the stone rows: in a putlog hole at the edge of the discontinuity a metal bar inserted longitudinally into the masonry can be seen. The wall discontinuity therefore came about during a construction phase and was not the result of structural failure, although it may have represented a wall area with weaker connections (Figure 4.38). In fact, there is no trace of

Figure 4.36 (a) In the transversal arches of the nave and in that of the crossing pillar, there are still traces of the anchoring of the temporary ties (green). (b) The metal connections of the temporary ties.

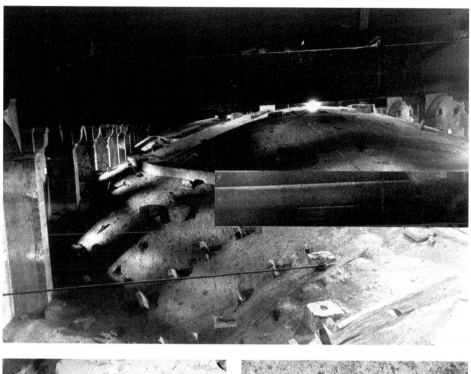

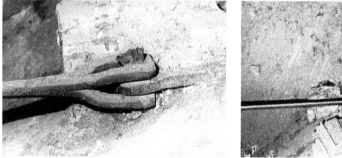

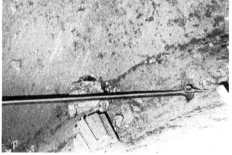

Figure 4.37 Garret of the nave showing the tie made up of three pieces inserted into the transversal arch. (below) Details of the ties.

it on the opposite side of the wall towards the nave, nor are there any visible repairs to the plasterwork; only higher up, beyond the gallery, is there a slight crack in the same position, as well as signs of plasterwork repairs. This crack and the discontinuity in the garret are now monitored by electronic crack-measuring devices.

The transversal structures corresponding to the flying buttresses differed in terms of the greater strength conferred to those with pillars on the outer flanks, at the end of the second bay. However, given the architectural structure of the building at this stage of the works, the marked difference between the transverse and longitudinal constructive arrangement of the basilican body was more significant. While the former had now reached conditions close to that of the completed building, for the latter, the situation was quite different due to strong

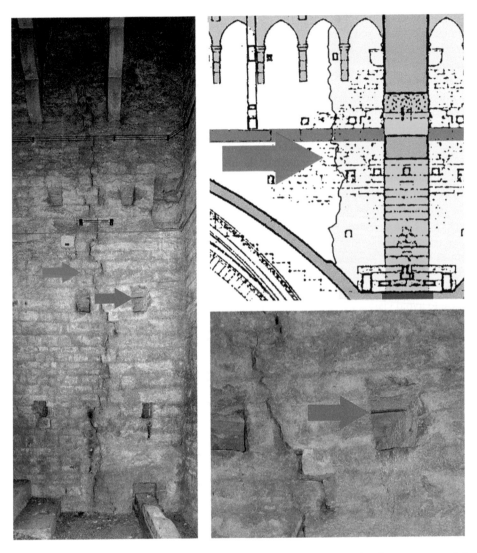

Figure 4.38 Garret of the north aisle: in the wall, the interruption in the masonry of the first two bays is clearly visible (red arrow); the drawing indicates the masonry of the first two bays (yellow), the flying buttress (green), the stone tie (blue) and the corbels (orange); a metal bar within the masonry can be seen in a putlog hole (blue arrow).

asymmetry in the buttressing system. In fact, on one side the two bays abutted the massive masonry of the façade, while on the opposite side, the building was interrupted at the second pillars and at the buttresses of the flanks, which were isolated wall elements that had still not been fully constructed.

In each aisle reinforcements and the transverse wall connections of the diaphragm arches could be relied on, but the aisles were two interrupted parts of the building subject to the thrusts of the vaults of the nave.

In particular, all this posed a problem linked to the lack of stability of the pillars on the interrupted side. So during the construction of the cathedral, after the second vault of the nave had been built and even before the centring was taken down, despite all the precautions adopted, a serious static problem developed that required new important decisions to be made.

In July and August of 1366, the Opera approved numerous resolutions related to these difficulties. The decisions appeared to be chaotic and apparently contradictory. On 13 July, it was decided to interrupt the work in the first two bays in order to define the apsidal area, at the same time specifying that the basilican body would be made up of four bays. A few days later, they addressed the problem of the thin cracks (the "peli") that had appeared in the vaults of the aisles. The solution to strengthen and repair the aisles was particularly complex and, as a first step, more than once it was decided not to dismantle the centring of the vaults of the nave so as not to increase the amount of thrust (Guasti, 1887, docs. 141, 143–145, 147).

The salient choice of the experts, who were consulted several times, was to define the design of the apsidal area, which was developed in a very short time and approved on 13 August (Guasti, 1887, doc. 150). Appreciation for the design also involved structural evaluations, in that it was considered "very strong" also due to the heights that would have to be reached without resorting to visible ties. It is significant that a similar intention was expressed in conjunction with the instabilities occurring in the first two bays under construction: the aim was still to avoid visible ties, using temporary ties also in the arches of the crossing.

The builders were well aware of the important role the apsidal area would play in terms of the stability of the basilican body. Thus, it was pointed out "che il decto disengno vada inanzi, e secondo quello si proceda a perfetione della decta chiesa" [that said design should be pursued, and on the basis of it said church should be completed]. This last statement specified the link between the construction of the apsidal area and the anchoring, on this side, both of the flanks of the aisles and of the longitudinal arches. This would have made it possible to obtain equivalent conditions of stability at both ends of the basilican body.

In addition to the design of the "fourth columns" attaching to the crossing, the dimensions of the nave pillars were altered: starting with the third archway, there is an increase in the dimensions of all their faces, and although the change is subtle (the total size at the base goes from 291 to 303 cm), the increase in the section of the shaft is almost 10%; identical dimensions can be found in the semi-pillars and in the framework of the "fourth columns" corresponding to the faces of the pillar.

The masters, instructed by the Opera to check the structures of the first two bays, found that the walls and pillars had maintained their verticality (Guasti, 1887, doc. 149). The laser scanner survey allows us to check the verticality of the structures at the edge of the second bay. At the height of the capital supports, the pillars have an external inclination of 3/4 cm for a total divergence of 7 cm. The semi-pillars lean towards the outside by 8/6 cm. This indicates that the end part of the aisles, on both sides, was slightly inclined towards the outside. The extent of the movements is limited and may justify the fact that the masters did not report it as this condition was not considered to present any particular structural danger. In any case, the deformation must have contributed to the formation and spread of cracks in the vaults.

The same masters recommended strengthening the first two bays with ties. It was considered appropriate to place a tie seven braccia below the springing of the two transverse arches of the nave and to continue it in the garrets of the aisles so it could be anchored to the external flanks (Guasti, 1887, docs. 149, 153). This would have represented a significant rethinking compared to the original intentions to place just temporary ties in the nave, which were then to be removed when the work was complete. In each of the four smaller vaults, two ties would have been placed "in the flanks" for a total of eight tie rods.

Today, the ties envisaged for the vaults of the aisles cannot be identified, as evidently they were not inserted. It is in fact unlikely that the actual ties, made up of two pieces, were partially inserted by demolishing the brick vaults that showed worrying cracks; moreover, the anchoring on the inner side would have been just as problematic as it corresponded to the area between the top of the keystone of the longitudinal arches and the cornice at the springer of the corbels of the walkway of the nave; this condition would have made it particularly complex to conceal the anchor bar without leaving any trace of it. Even the ties planned in the nave would have had to have been inserted by breaking through the wall just below the capitals on which the groin vaults rest; an operation of this kind must have seemed so reckless that they decided against it. The insertion of the ties would have had a strong impact on structures that were already affected by major instabilities.

The definition of the crossing plan led to different solutions if on 19 November 1367 it was decided to follow the masonry model in the Opera building near the Bell Tower, while all other projects produced in brick, wood or on paper must have been destroyed (Guasti, 1887, 207).

The representation in the "Triumphant Church" fresco of a church building that showed such evident affinity with the events of Santa Maria del Fiore under construction that it can be considered a model for its completion, fits this context (Rocchi, 2004, 93–95). This painting is part of a cycle of frescoes painted in the "Spanish Chapel" in Santa Maria Novella. Andrea di Buonaiuto was assigned this commission on 30 December 1365 and completed it in 1367 (Figure 4.39).

The painting of the "Triumphant Church" was produced in the period in which the first two bays of the basilican body were completed and the construction of the large pillars

Figure 4.39 "Spanish Chapel" in Santa Maria Novella, Florence; detail of the "Triumphant Church", frescoed by Andrea di Bonaiuto in 1366/1367.

of the crossing had started. The basilican body coherently has four large bays, with flying buttresses above the aisles, simplified with a single archway; in the flanks of the aisles and of the nave, each span has a trifora which instead does not reflect any real element: the nave is lit by oculi, and in each of the first two bays of the flanks, there are three trifora. The problem of the characteristics of the openings that light up the nave, consisting of the alternative of windows and oculi, is persistent and was raised again on 31 May 1367 when the vaults of the first two spans were built (Guasti, 1887, doc. 170). They seem to have considered a rethinking of the oculi created to illuminate the first two large vaults. However, the creation of rustic openings in the position of the nave openings could also be hypothesized, postponing the choice of whether to insert an oculus or a window to a later date.

The depiction of the octagonal crossing with the apses abutting it has a general layout that overall reflects the project that was started (Rocchi, 2004, 93–95). In the painting, it is interesting to note the placement of the windows between the half-domes of the apses and at the same level as those of the clerestory. The position of the windows might have been one of the aspects under discussion among the different projects produced for the apsidal area; in fact, they were then abandoned in favour of an oculus on each side of the octagon at a level higher than the half domes. Another aspect that was likely discussed, but that did not materialize, might have been the solution of the gable crowning the sides of the apses. In the "Triumphant Church" painting, Andrea di Buonaiuto depicted real elements, executed with inevitable simplifications, integrated with architectural inventions that likely reflect solutions that were debated at the time.

Having defined the configuration of the crossing, in October 1366, work continued with the creation of the third pair of pillars; in early January of the following year, it was decided to place the coats of arms of the Church and of King Robert [of Anjou] in the keystones of the longitudinal arches of this bay (Guasti, 1887, docs. 160, 162). The following August, it was decided to start constructing the massive pillars at the end of the basilican body, on which the dome would then rest. Francesco Talenti continued to be mentioned as the master builder along with Giovanni di Lapo Ghini. This document did not fail to emphasize the importance of the connection with these pillars for the overall stability of the aisles; indeed, it stresses that the construction of the vaults of the aisles could therefore continue in safety. A further document reiterated the importance of the fourth pillars in order to continue the work on the smaller vaults in safety (Guasti, 1887, doc. 189).

In August 1376, work was done on the north longitudinal arch, which abuts the pillar of the crossing (Guasti, 1887, doc. 256). This document also provides interesting information on the finishing of the stone ashlars of the arch, for which there were recommendations to work the surface with a point so that the plaster would stick better; this confirms that the current visible facing of the arches, with its rustic surface unlike the pillars and pilasters worked with a toothed hammer, is the result of restorations. Meanwhile, work had been carried out on the vaults of the third bay of the aisles since the previous summer. The construction of the "large pillars" of the crossing, with a large barrel vault connecting them, continued in April 1377 (Guasti, 1887, docs. 244, 247, 269). On the crossing side, at the end of the longitudinal arches of the basilican body there were massive supporting structures, in total respectively 4.80 m wide and 8.44 m long.

Once the buttressing had been put in place on both sides of the basilican body, and safe conditions had therefore been achieved a decade after the structural crisis of 1366, it was not surprising that the most demanding construction site operations regained speed. On 31

December 1378, the entire vaulted system of the basilican body was finished with the completion of the fourth vault of the nave (Guasti, 1887, docs. 286, 294); the two groin vaults of the nave were built in less than a year and a half.

In the fourth groin vault of the nave, as the masonry structure of the extrados is visible, a distinct constructive solution can be seen, probably adopted due to the considerable extension of the brick vaults. To avoid problems of direct interlocking, the webs were made by overlapping them on the edge of the diagonal ribs in "Pietraforte" stone of the intrados; between the webs and the diagonals a sort of formwork of irregular width (about 35/40 cm) was created, where they built the arches which could be more easily interlocked with the skewed and necessarily irregular structure of the adjacent webs (Figure 4.40). It would seem that due to the unusual size of the vaults, they feared a possible fracture line might appear on the diagonals if a more traditional herringbone connection between webs was used.

Further information on the worksite practices can be obtained from some metal elements found at the intrados of the diagonal ribs of the vaults of the nave. These are metal plates placed in the joints of the ashlars, which were probably flat irons used to anchor the centring of the diagonals, a criterion reminiscent of the centring anchors at the corners of the intrados of the dome. Other metal elements with a probable similar function can be found at the intrados of the transverse arches of the nave and of the arch of the drum towards the façade. In some cases, they were flat irons bent on the ashlars, and other times, they were cut and can be found in the joints.

The construction of the longitudinal and transverse arches of the nave and aisles and of the east arch of the drum left traces of the support slots of the centring on the abacuses of the capitals. In the first case, there are usually recesses to support two paired beams, and in the second case, the number of beams increases up to five, with recesses also in the cornice supporting the corbels of the gallery. In the large arch, putlog holes were also added, which suggest centring with a more complex framework including beams and struts on several levels (Figure 4.41).

The same constructive protections used in the first two bays were essentially replicated in the third and fourth bays: ties in the aisles, the diaphragm arch system and flying buttress with a tie below the three salient arches of the latter, ties inserted into the centre of the vaults of the aisles and temporary ties in the transverse arch of the nave. In the flanks of the nave incorporated into the garrets of the aisles, in these bays, there are no pilasters corresponding to those of the flanks, as they were abandoned following the layout of the aisles decided in the summer of 1357. So the pillar of the flying buttress abutting the inner flank was in this case built entirely with squared ashlars and given symmetrical winged pilasters, assuming an overall width equal to the buttress above of the flank at the level of the oculi. But the pillars of the flying buttresses of the third archway were smaller.

It was decided not to use flying buttresses on the side adjacent to the drum. Alternatively, they created masonry overhanging the vaults of the aisles, conceived as a sort of buttress on which rested the springer of the groin vault of the nave (Figure 4.42). It must be remembered that this area represents the transition from the basilican body to the domed area, namely the connection that may have been subject to possible subsidence caused by the weight of the dome, which in fact occurred, as we shall see in the following chapter. In a similar context, they used overhanging buttresses directly connected to the pillars of the crossing, thereby creating a support that was certainly more stable than the slender structures of the flying buttresses.

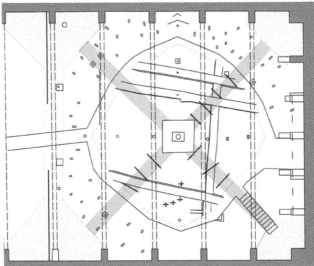

Figure 4.40 Nave, plan of the vault adjacent to the crossing; the webs of the crossing are interposed with diagonal arches in brick (grey) visible at the extrados (above).

Figure 4.41 (a) In the arch between the nave and the crossing, there are visible anchors for the temporary ties (green circles), made up of double eyelets (images on the right). (b) The support slots of the centring used in the construction of the longitudinal arches are also highlighted (red and purple).

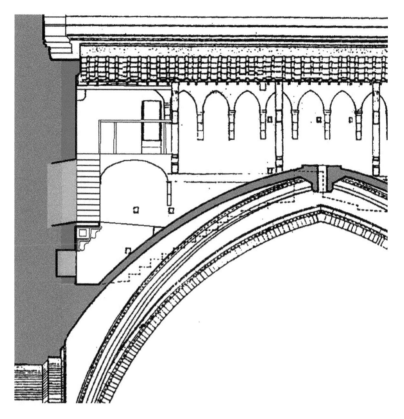

Figure 4.42 Close to the pillars of the crossing, overhanging masonry (red) was used to counteract the thrusts of the vault of the nave.

Near the top of the transverse arch placed at the end of the nave, a transverse tie was proposed again, which is now completely incorporated into the adjacent drum, except for a short section. This tie testifies that this part of the drum, beyond the level of the frontal arches of the domed space, was built after the completion of the fourth vault of the nave: as the walls at the level of the oculi of the nave were still not attached to the drum, they continued to be stabilized through the use of ties.

In the large arches of the crossing, actual barrel vaults between the pillars of the dome, they continued to set up connections for temporary ties, two for each arch, consisting of Y-shaped eyelets in special slots, where the eyelet of the temporary tie would have been introduced and then blocked with a transverse dowel. In the four archways, in addition to traces of the dowel slots used to block the connections between the eyelets, there are also recesses above the same eyelets which were necessary to insert and remove the temporary ties.

The attention paid to the relationship between the basilican body and the domed space can also be seen in the particular care taken in the masonry structure of the arch on which the nave rests: in the springer area towards the inside of the octagon, there are through stones between the abutment wall and the archivolt of the arch. To this end, they used continuous ashlars characterized by stretches with different inclinations in the transition from the wall to the arch.

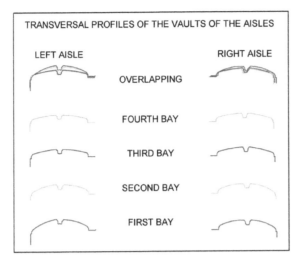

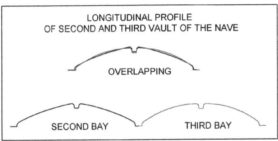

Figure 4.43 Variations of the vault profiles going from the first two bays to the subsequent ones; (a) the vaults of the aisles were made less rampant; (b) in the nave, the vaults were shaped like the arc of a circle.

With the construction of the third and fourth bays, with respect to the two previous ones, some changes were made to the profiles of the vaults. The laser scanner survey of the aisles highlighted that the external transverse webs of the aisles had a less marked rampant profile, with the intent of creating more rigid vaults more suited to shore up the thrusts of the nave. In the larger vaults, the directrices had a more marked curvature, assuming the overall appearance of the arc of a circle in the longitudinal and transverse axes (Figure 4.43).

4.5 The basilican body in Brunelleschi's time

After the difficulties experienced at the construction site after the first two bays of the aisles had been built, a subsequent dangerous situation for the basilican body developed during the construction of the dome. Mention of it was first made in a resolution passed in January 1431, which referred to uncertainties about the structural strength of the cathedral aisles, which threatened to collapse (Gli Anni, 2015, II 2 1 c. 136).

First, it should be taken into account that according to a date engraved into the plaster which could be read as 1429 or 1430 (Haines, 2013, 57–58), by this point, the dome had reached the level of the third walkway; this means that its wall mass was well over half its overall weight (Figure 4.44), and the loads on the pillars and the thrusts transferred to the adjacent structures were close to those of the completed dome. Just how worrying the situation was that had arisen was also demonstrated by the fact that work to construct the dome was interrupted and resumed at a steady pace in the second half of 1431 (Haines, 2013, 59).

The bay adjacent to the drum still reveals the effects of the interaction between the basilican body and the pillars of the dome, resulting from the subsidence of the latter at the base. A running crack formed in the flanks of the central bay, extending from the longitudinal arches to the walls and oculus above.

The instability created more pronounced damage on the north side. The crack that formed diagonally can still be seen in the archivolt of the arch; the central part of the fracture was repaired with plasterwork and two metal cramp irons, one placed radially like the ashlars and the other roughly horizontally. The fracture also created instability at the intrados and extrados, which was corrected in the former case by tessellating some ashlars, and in the latter, it was repaired with mortar.

The crack is somewhat visible also in the wall above the arch; this would seem to indicate that the interconnection between the aisles and the dome still gives rise to cracks today. The effects produced in the Talentian gallery are however much more pronounced. The cornice

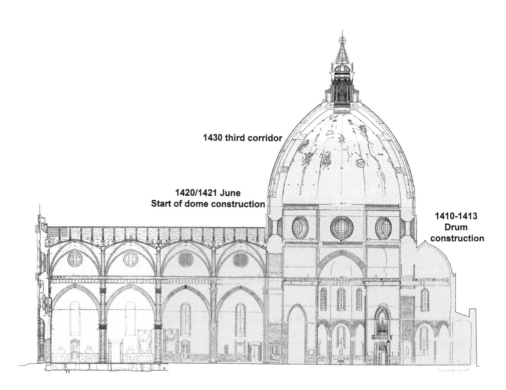

Figure 4.44 In the longitudinal section, some main construction phases are indicated, the first two bays (yellow), the subsequent completion works up to the apses (green), the drum with oculi (purple), the dome up to the third corridor (light blue).

supporting the corbels on which the walkway rests has an abrupt lowering strengthened by two metal brackets; the subsidence in the handrail of the gallery was concealed by inserting considerable layers of mortar below it. Despite this workaround, in the central part of the bay, the handrail continued to lower by over 6 cm in just a few metres (Figure 4.45).

The effects produced at the level of the oculus are particularly visible from the outside. A marked disruption in the horizontal white marble and serpentine bands was created under the circular opening, which was repaired with abrupt skewed connections. In the part above the

Figure 4.45 Traces of instability in the bay adjacent to the crossing: reparation of the masonry of the longitudinal arch and of the cornice at the base of the walkway; the diagram shows the subsidence of the bay measured on the handrail of the walkway.

oculus, at the base of the cymatium, the widespread detachment of serpentine tesserae from the carving in the solid white marble can be seen (Figure 4.46).

In the unstable area, where the restoration work does not all seem to be coeval, tie-beams were also used in two trusses of the garret of the north aisle in order to transversely connect the two longitudinal walls of the north bay adjacent to the drum; the tie-beams of the trusses were connected to an anchor bar and are visible on the inner side between the corbels of the gallery.

The effects of the instability caused by the dome left fewer traces in the south arch of the fourth bay. However, above the keystone, there is a clear lowering of the cornice supporting the corbels of the gallery.

The consolidation of the marble ribs of the dome confirms greater subsidence of the pillar of the drum on the north side of the nave: only those of the north-west side have external stirrups, which on one rib go as far as the third walkway.

The aforementioned resolution of 1431 arranged for the first consolidation works, but just a few weeks later, the first changes to the solution for the reinforcements emerged.

Figure 4.46 At the level of the oculi, the marble incrustations show detachment and skewed connections in the horizontal cornices.

The resolution of 26 January refers to a complex system of reinforcements, with double iron and wooden ties to be inserted into the aisles; additional ties were planned for the larger vault, again double and made of iron and wood, to be placed at an intermediate height and at the top of the transverse arches. Filippo Brunelleschi, Lorenzo Bartoli and Battista d'Antonio were involved, but the creator of the model was Brunelleschi.

In the subsequent resolution dated 7 February, it was specified that the tie to be placed in the arches of the nave would have been more effective if it had been extended to the flanks of the aisles, as a single continuous reinforcement system placed at the same level; to this end, the tie of the nave had to be lowered by four braccia (Saalman, 1980, 270, doc. 253.6).

Two metal ties were placed next to each pilaster in the nave, which continued with wooden ties into the garret of the aisles. In order to understand how they decided upon the solution for the reinforcements which was then put in place, it is useful to consider some subsequent documents and correlate them to the work that was actually done; the arrangement of the actual ties of the nave, which flank the two pilasters, were gradually developed, and other variations were necessary (Figure 4.47).

The following 27 February, Brunelleschi was instructed to commission Bernardo di Zanobi to produce the iron for six ties necessary to strengthen the church; two ties were needed to flank each of the three pilasters of the nave. After a few days, on 2 March, the carpenter Antonio di Melato made the timber model of the ties for Brunelleschi (Gli Anni, 2015, II 2 1, c. 137v, II 4 13, c. 7v), concerning the wooden trestled framework to be placed at the foot of the pillars of the flying buttresses.

In March and June, the "master stonecutter" Jacopo di Sandro provided the timber for the ties that were to be placed in the church. The following month, the masters Piero di Tuccio, Antonio di Lionardo, Leonardo di Giovanni, Francescho di Giovanni Targia, Biagio d'Angniolo, Gherardo di Nicholaio and Filippo di Giovanni were paid for further consultations on the ties (Gli Anni, 2015, II 4 13, c. 9, c. 12v, II 4 13, c. 15v).

The purchase of iron for ties was recorded several times in the subsequent months (Giorgi & Matracchi, 2006b, 323, notes 84–85), and the resolutions of 6 June and 25 September 1432 specified that they were ties for the nave (Gli Anni, 2015, II 4 13, c. 33v, c. 42).

The quality of the iron was also checked in the meantime. On 17 June 1432, an order was given to return to Simone di Michele the iron supplied for the ties as it had been found to be defective (Gli Anni, 2015, II 2 1, c. 162).

While the wooden and iron ties were being produced, after consultations with Brunelleschi, it was decided in April 1433 to eliminate the two temporary ties of the arch of the crossing placed in continuity with the nave; the main altar had temporarily been placed below this arch pending completion of the construction of the dome (Gli Anni, 2015, II 2 1, c. 196v, II 4 14, c. 20). The ties were not deemed suitable for providing stability; moreover, perhaps referring to the rest of the arches of the crossing which did not have ties, the ties of this arch were thought to give the church a "rustic" feel. A few days later, the stonemason Checco d'Andrea Fraschetta cut the two ties using chisels (Gli Anni, 2015, II 4 13, c. 54v).

The first tie had still not been put in place since the first initiatives of January 1431. It was not until 19 May 1433 that Brunelleschi and the master builder Battista d'Antonio were instructed to oversee the installation of the ties (Gli Anni, 2015, II 2 1, c. 198v). But in the following months, work to complete the ties continued at the same time, which included the welding of the ties, the supply of elms for the wooden ties at the base of the flying buttresses and the purchase of the anchor bar (Giorgi & Matracchi, 2006b, 323, notes 92–94).

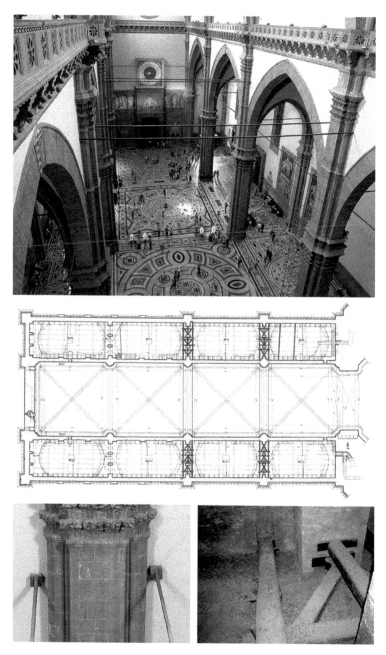

Figure 4.47 (a) View of the nave with the added couples of ties. (b) Plan of the level of the garrets of the aisles; the image indicates the ties partially inserted into the vaults and those of the flying buttresses (red), the wooden ties (brown) anchored at the base of the two flying buttresses and those added in the north-east bay; the metal ties that cross the nave and continue with wooden elements at the foot of the flying buttresses. (c) Details of the metal ties in the nave and of the wooden ties in the garret.

The document that attests the installation of the first tie, under the guidance of Brunelleschi and the master builder, dates back to March 1434 (Gli Anni, 2015, II 2 1, c. 211v). The three years that had elapsed since the resolution of 1431 seem to indicate that the builders had wished to monitor the static behaviour of the structure over time with the hope of avoiding having to insert visible ties into the nave. Since 1366, when damage occurred to the vaults of the aisles, they procrastinated for almost seventy years before the visible ties in the nave were considered indispensable.

The aforementioned document dated 1434 is very important as it allows us to understand that the tie of the central pillar of the aisles had been created and that the addition of two intermediate arches of the flying buttresses pillars was down to Brunelleschi (Giorgi & Marino, 1988, 56). He created this doubling on two levels of the flying buttresses, shifting a part of the thrusts from the nave to a lower level on the flanks of the church, where the containment action of the ties inserted at the base of the same flying buttresses could also be relied on (Figure 4.48).

It is not surprising that the first double tie to be placed was that of the central transverse arch of the basilican body, where there was most concern about deformation caused by the action of the thrusts of the large vaults, and only here was the doubling of the flying buttresses carried out (Figure 4.49).

In the nave, the tie was made of two paired metal tie-rods positioned at the sides of the semi-pillars and connected to wooden ties that continued into the garrets of the aisles. It should be noted that these ties were placed well below the springer of the groin vaults of the nave, as their purpose was not to counteract the thrusts of the large groin vaults of the nave – which would be handled by the flying arches – but to ensure the overall stability of the basilican body after the subsidence of the pillars of the dome adjacent to the same aisles (Figure 4.47).

The static behaviour of the central section investigated, investigated using numerical models, took into consideration the structural system formed by the central arch, the flying buttresses and the double ties with wooden anchoring trestled framework, as well as the basilican body without paired ties (Miceli & Papi, 2004, 312). The result obtained shows that the ties, positioned well below the springer of the vaults, changed the static behaviour of the structures, overall resulting in a significant reduction of the tensile stresses in the areas of the diaphragm arches and flying buttresses.

The purchase of iron documented from May 1434 can be linked to the other double tie inserted between the third and the fourth bay (Gli Anni, 2015, II 2 1, c. 216v). The works referring to this tie are recorded up to 1437 (Saalman, 1980, 270, doc. 253.23/24/25).

The transversal ties inserted into the basilican body required a complex system of connections in order to join the metal ties to the wooden ties and to make the latter part of a wooden frame anchored at the base of the pillars of the flying buttresses. Each metal tie of the nave had an eyelet which was anchored with a dowel to a double Y-eyelet on a stretch of a metal tie inserted into the breach of the wall to the side of the pilasters. This short stretch of tie extended as far as overlapping the side of the wooden ties, with an average section of 30×33 cm, which continued into the garret of the aisles. The connection between the metal and wooden ties was created with transversal dowels that completely cross the wooden tie and are anchored by pins to a metal bar with a square section with sides measuring 8 cm (Figure 4.48). The wooden ties, in addition to continuing the metal ones of the nave, are part of a wooden trellis that uses transverse and diagonal elements to anchor to the base of the pillars of the flying buttresses. The transverse elements, using specifically created slots, cross the beams

Figure 4.48 (a) The north flying buttress, positioned in the centre of the aisle and (b) detail of the anchoring system for the wooden ties.

and are fixed by wooden wedges. Several transversal holes were therefore created in the beams placed in continuity with the metal chains of the nave for the anchoring of the same ties and the wooden trellis. Perhaps, this is why an additional beam, with an average section of 18 × 30 cm, was added above the two beams, fixed to them with nails and metal brackets. The upper beam

Figure 4.49 Section corresponding to the central pillars of the aisles; the entire transverse structure has been highlighted, with the metal ties (red) and the wooden ties (brown); the latter, which continue in the nave with metal ties, were added by Brunelleschi.

integrates the trellis structure with additional transverse wooden elements placed to counteract the central pillars of the flying buttresses (Giorgi & Marino, 1988, 57). The outer ends of the wooden ties are anchored to the outer walls of the church using an anchor bar (Figure 4.50).

Figure 4.50 Detail of the section and plan of the south flying buttress, positioned in the centre of the nave highlighting the metal ties (red), the wooden ties (green), the segmental arches (orange) intermediate to the pillars of the flying buttresses.

In the garrets of the north aisle, a metal tie was added on both sides of the first flying buttress. The one made up of a square iron with an anchor concealed by the marble cladding of the flank may be from the 14th century, and in any case, it dates to after 1366; the other, with a circular section, made up of two pieces with an eyelet connection resting on a small pillar and an anchor bar adjoining the cladding of the flank, would seem to be part of a restoration.

The vault of the nave adjacent to the drum underwent a remarkable intervention, still related to the consolidation of the basilican body in Brunelleschi's time.

The extrados of the vault was abutted by three arches, for the most part parallel, and a fourth arch, approximately orthogonal to them, which is interrupted where it intersects with these arches. The system of four extradosal arches, about 60 cm wide and 30 cm high, is rotated with respect to both the axes of the church and the diagonals; moreover, its extension is dictated by the extradosal offsets of the groin vault, where the reinforcement arches abut.

Metal brackets were added to the diagonal ribs, which were fixed to the pietraforte ribs on the intrados, while on the extrados the brackets were connected to an anchor bar placed on the diagonal brick arches between the webs, or on the reinforcement arches where there was an overlap between diagonal arcs and the added arches. Longitudinal metal elements encased in brackets were also placed on the intrados of the diagonal stone ribs: this secured all the ashlars in the area reinforced by the stirrup (Figure 4.51).

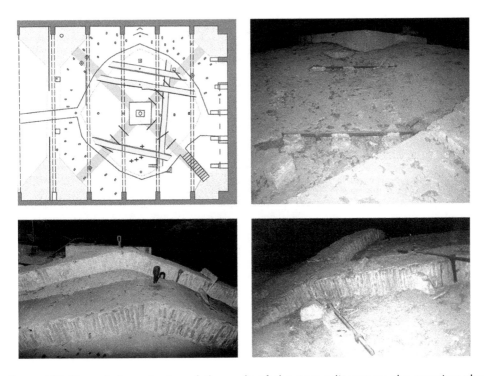

Figure 4.51 Plan of the extrados of the vault of the nave adjacent to the crossing; the works to reinforce the vault involved the addition of extrados arches (yellow) and brackets that connect the diagonal brick arches to those in stone of the intrados (top right); metal brackets were also added in other diagonal ribs, resting them directly on the extrados or interposing a wooden element (bottom).

Two extrados arches were placed transversally to the axes of the web adjacent to the north side, where subsidence had produced the greatest instability. There was just one arch on the opposite web. The fourth arch reinforced the web adjacent to the drum, extending for a stretch to the adjacent ones. Each of the diagonals delimiting the web connected to the drum had four stirrups; a similar solution was used in the rib between the south and west webs, while a single bracket was inserted into the remaining rib. In the aisles, all the ribs were reinforced with brackets, except in the third and fourth north bays (Figure 4.52).

The intrados of the vaults provides an overall view of the consolidation of the diagonal ribs, carried out in different eras and always using metal brackets, but with a smaller span than the fourth large bay. In the nave, the interventions involved the second and third vaults (Figure 4.53).

Figure 4.52 Extrados of two aisle vaults with the reinforcement system of their diagonal ribs.

Figure 4.53 View of the intrados of the vaults with the diagonal rib areas (red) consolidated by brackets.

The study of the state of tension of the ties of the basilican body seems to reflect, in some tensile stress values, the special circumstances faced during the construction phases of the building (Blasi & Giordano, 1991, 25–27). The ties of the aisles subject to the greatest stress are the transverse ones of the first two bays, which had to address the structural crisis of 1366 that occurred after the vaults of the nave had been constructed. In the ties almost at the top of the flanks of the nave, inserted into the keystone area of the transverse arches, the tensile stresses are different; the tensile stress of the tie positioned between the two bays is similar to that of the transverse ties of the aisles, while it reduces in the tie of the arch at the end of the first two bays.

The tensile stress of the longitudinal ties of the first two bays is less indicative, in that those anchored on the façade were replaced; there is only a chronological reference for the one on the north, as it was inserted after the earthquake of 1895. The south longitudinal tie of the fourth bay is not original either, as it was inserted in 1985. It is not known how long these ties were lacking; nevertheless, this static condition did not compromise the stability of the building as the longitudinal arches constituted a system of diaphragm arches with four archways, flanked on one side by the pillars of the dome on the other by the façade wall.

For an unspecified period, the longitudinal arches to the south were found without ties in the first and fourth bays; the thrusts of the two lateral arches that had no ties may have produced local effects, reducing the tensile stresses in the ties of the two central bays, which are not as high as those recorded in the longitudinal ties.

In the paired ties, inserted in Brunelleschi's time, considerably lower stresses were estimated than those of the transverse ties of the first two bays of the aisles; moreover, between the two double ties, that of the central arcade is subject to greater stress, as might have been expected; here, in fact, there are flying buttresses to which lower arches were added between the pillars, thereby partly transmitting the thrusts of the vaults of the nave to the end of the Brunelleschian tie. The low tensile stress of the double ties can mainly be attributed to two factors: these were ties added not due to the static shortcomings of the constructive structure, based on the containment of transverse thrusts by means of flying buttresses, but due to the subsidence caused by the weight of the dome; moreover, as the 15th-century ties were double, overall, their resistant sections are larger than those of the corresponding ties of the aisles.

4.6 The construction paradigm shift in Santa Maria del Fiore

In the cathedrals of Pisa and Siena, the aim was to achieve continuity of the architectural space between the nave and the crossing, following a vision that led right up to the apse. In the Pisan church, there is a transition from the large granite columns in the aisles to masonry pillars at the corners of the rectangular crossing, which is cross shaped with symmetrical dimensions. The Sienese church reinforces the longitudinal continuity by reusing lobed pillars with duotone stone rows in the crossing; the pillars have a larger masonry section and are enriched with the multiplication of dihedrons between one lobe and another. In both churches, the arch between the nave and the crossing is higher than the arch leading into the apse.

The crossings of these cathedrals have a longitudinal extension equivalent to the width of the wings of the transept. It should be taken into account that in the case of Siena, the crossing was located in the transept-apse area of the intermediate cathedral defined in the 1260s, which was then completely dismantled following changes. The major innovation accomplished in Siena cathedral with the hexagonal crossing should be pointed out, delineated by pillars placed in continuity with the nave and by pillars shifted towards the wings of the transept, achieving intermediate alignment with the aisles.

The design of the Duomo Nuovo of Siena (1339), which was considerably larger than the earlier cathedral and arranged orthogonally to it, intended to adapt the crossing of the earlier building by eliminating two pillars from the hexagonal plan which would have been in line with the nave of the new church. The sequence of events regarding the Duomo Nuovo, ending with the collapse of most of the structures that had been erected and the abandonment of this project, created the conditions that led to the definition of the current basilican body of Santa Maria del Fiore. Francesco Talenti and Benci di Cione, who were called upon as consultants in the spring of 1357 to give an opinion on the critical conditions of the Duomo Nuovo and then worked for a long time on the construction of the aisles of Santa Maria del Fiore, created a close connection between the two construction sites. Their analysis of the vulnerabilities of the Sienese church must have provided them with sufficient understanding to come up with the necessary constructive solutions at the Florentine site, where in fact work had dwindled for a long time and where, from the summer of 1357, they started to build vaulted aisles that were significantly larger than those of the Duomo Nuovo, creating roughly square bays of unusual width in the nave.

Only the aisles were fitted with ties. The transversal structures placed between the bays were reinforced by diaphragm arches with flying buttresses above, concealed in the garrets of the aisles. This solution recalled the flying buttresses hidden in the garrets of Pisa cathedral, originally visible from the matronea, which served to counteract the thrusts of the transverse arches of the crossing. In Santa Maria del Fiore, the aisles and nave had brick groin vaults of a similar thickness, one-and-a-half bricks thick in the initial parts closest to the springer and one brick thick in the remaining parts. But, in this regard, it should be pointed out that the different width and position of these vaults had been carefully considered; in the nave, the part that is one-and-a-half bricks thick extended much further than the part that was one brick thick.

A comparison of the plans for the piers of the aisles in the cathedrals of Pisa (considering those flanking the nave), Siena (including the surviving pillars of the Duomo Nuovo) and Florence reveals interesting dimensional variations. As could be expected, the granite columns of Pisa cathedral, covered by trusses, have the smallest section. In the vaulted aisles of Siena cathedral, the section increase of around 30% recorded in the transition from the pillars of the aisles defined with the intermediate cathedral to the pillars of the Duomo Nuovo appears to be moderate and might have been one of the joint causes that led to the collapse. In Florence cathedral, the increase in the resistant section of the pillars is, respectively, of 242% and 263% much bigger with respect to those of the Duomo Nuovo (Figure 4.54).

There is no specific data on the characteristics of the inner core of the aisle pillars of Siena cathedral. However, they could reasonably be considered similar to those of the pillars of the hexagonal crossing, which were examined by core drilling and micro-perforations for endoscopic inspections (Tosti, 2006). It emerged that the stone facing is connected to an interior wall of bricks laid on good quality mortar. Information on the internal masonry characteristics of the pillars of Santa Maria del Fiore is lacking. However, considering the chronological proximity of execution and the standard organization of the construction site under the control of the Opera, it could be assumed that they were constructed in the same way as the interrupted pillars at the top of Giotto's bell tower. In this case, the pietraforte

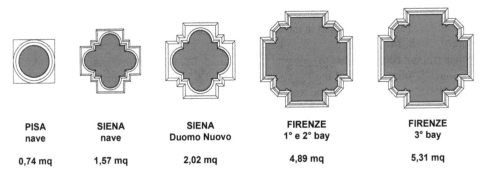

PISA	SIENA	SIENA	FIRENZE	FIRENZE
nave	nave	Duomo Nuovo	1° e 2° bay	3° bay
0,74 mq	1,57 mq	2,02 mq	4,89 mq	5,31 mq

Figure 4.54 Comparison between the sections of the pillars of the aisles of Pisa, Siena and Florence cathedrals; in the cathedral of Santa Maria del Fiore, the section of the pillars was increased by around 10% when the second two bays were constructed.

masonry of the facing, with ashlars of different depths, continues inside with rows of similar height, made up of roughly hewn stone positioned on beds of rather abundant mortar.

The structures of the basilican body of Santa Maria del Fiore can be considered rather slender overall, taking into account the size of the building: the walls above the longitudinal arches, with thicknesses that reduce with the vertical progression 100/90/74 cm, rise to over 43 m.

The constructive protections set up during the works, using metal, wooden and stone ties, played a fundamental role in the organization of the construction site of Santa Maria del Fiore. In the construction phases, they made a significant contribution to how the various static conditions encountered were dealt with as work progressed, which were very different from the structure of the church once the work was complete.

After the construction of the first two bays, the skill and understanding required to tackle the large size encouraged the builders to create a huge octagonal crossing flanked by a triconch; the distance between the opposite vertices of the octagon is on average 45.15 m, a width that exceeded the total width of the aisles and that then dictated the planimetric size of Brunelleschi's dome.

On the basilican body side, between the aisles and the crossing, connections were created in the pillars of the dome covered by a long barrel vault which was rather narrow in that its profile was determined by the transverse arches of the aisles. In order to connect with the octagonal plan of the crossing, this transition has a slanted ending reflecting a pragmatic geometric adaptation. These pillars of the dome were therefore divided into two walls, one aligned with the sides of the church and the other with the longitudinal arches. Sacristies were created in the two remaining pillars of the dome.

The domed space of Santa Maria del Fiore, compared to the crossings of Pisa and Siena cathedrals, is more independent from the basilican body. Imposing pillars were placed at its base, with the extension of the wall mass clearly prevailing over the character of an isolated pier made up of a base, shaft and capital. The walkway on corbels which extends from the nave to the crossing and the springer of the half domes of the triconch system was certainly not enough to create the strong architectural continuity between the aisles, crossing and apse, which dominates in buildings like the cathedrals of Pisa and Siena.

The distinct characteristics of the octagonal triconch space of Santa Maria del Fiore influenced Leonardo da Vinci, who was inspired to create a sketch of a church, placing the tribunes leaning against the crossing in the foreground and a drum surmounted by a dome above (Pedretti, 1988, 148–149).

It has still not been possible to gain any insights into the construction characteristics of the enormous masonry masses of the pillars of Santa Maria del Fiore, with thicknesses reaching 8 m. Such massive walls recall the Rognosa, Becci and Cugnanesi towers of San Gimignano (Giorgi & Matracchi, 2019), which have solid walls at the base rising up beyond the external ground level. These bases are made up of a stone facing, a casting of stone conglomerate and copious high-quality mortar. Similar precedents should be no great surprise as architectural innovation is inevitably combined with the persistence of construction traditions and construction site practices.

References

Aita, D., Barsotti, R., Bennati, S., Caroti, G., Piemonte, A. (2017). 3-Dimensional geometric survey and structural modelling of the dome of Pisa cathedral. *The International Archives of the Photogrammetry, Remote Sensing and Spatial Information Sciences, XLII-2/W3, 3D Virtual Reconstruction and Visualization of Complex Architectures*, 1–3 March 2017, Nafplio, Greece. DOI 10.5194/isprs-archives-XLII-2-W3-39-2017.

Bacci, P. (1917) Le fondazioni della facciata del secolo XI nel Duomo di Pisa. *Il Marzocco*, XXII, n. 35, 2.

Bagnoli, A. (2003) Alle origini della pittura senese. Prime osservazioni sul ciclo dei dipinti murali. In: Guerrini, R. (ed.), *Sotto il duomo di Siena. Scoperte archeologiche, architettoniche e figurative*. Milano, Silvana Editoriale, 107–151.

Bartoli, G., Betti, M. & Orlando, M. (2013) *Cupole medievali. Il duomo di Siena. La diagnostica strutturale per il cantiere di restauro / Medieval Domes, the Cathedral of Siena, Structural diagnostics for restoration works, II*. Firenze, Alinea.

Bebber, A. E. (1988) Datazione dendrocronologica di travi della copertura della navata maggiore. In: Rocchi, G. et al., *S. Maria del Fiore: Rilievi, documenti, indagini strumentali. Interpretazione. Il corpo basilicale*, 61–62.

Belli, G. (1994) La cattedrale di Santa Reparata. In: Gurrieri, F. et al., *La Cattedrale di Santa Maria del Fiore*, I. Firenze, Giunti, 15–34.

Benucci, V. (1986) *Siena città dei terremoti? Rispondono la storia, la letteratura e la scienza*. Poggibonsi.

Bevilacqua, M. (2015) *I progetti per la facciata di Santa Maria del Fiore (1585–1645). Architettura a Firenze tra rinascimento e barocco*. Firenze, Olschki.

Blasi, C. & Giordano, S. (1991) Rilievo dello stato di sollecitazione del sistema catenario della chiesa di Santa Maria del Fiore. *Bollettino degli Ingegneri*, gennaio-febbraio 1991, n. 1–2, 25–27.

Borgherini, M. (2001) *Disegno e progetto nel cantiere medievale. Esempi toscani del XIV secolo*. Venezia, Marsilio.

Butzek, M. (ed.) (1996) *Il duomo di Siena al tempo di Alessandro VII : carteggio e disegni (1658–1667)*. Munchen, Bruckmann.

Butzek, M. (2006) Chronologie. In: Haas, W. & von Winterfeld, D., *Der Dom S. Maria Assunta. Architectur, 3.1.1.1.* Munchen, Deutsche Kunstverlag.

Butzek, M. (2010) Il Duomo di Siena: ipotesi su alcuni documenti del Duecento e sulle vicende costruttive della parte orientale della chiesa, *Prospettiva. Rivista di storia dell'arte antica e moderna*, 139–140, Luglio-Ottobre 2010, 111–112.

Caldano, S. (2018) L'attività di Francesco Talenti alla cattedrale di Firenze e in altri cantieri centroitaliani: primi risultati di ricerca. *Studi e ricerche di storia dell'architettura*, n.4-2018, 78–85. DOI 10.17401/STUDIERICERCHE-4/2018-caldano

Calderoni Masetti, A.R. (2014) Restauri ottocenteschi alla facciata del duomo di Pisa. In: Calderoni Masetti, A.R., *Arti in dialogo. Studi e ricerche sul Duomo di Pisa*, Modena, Panini, 40.

Carbonai, F., Gaggio, G. & Salmi, M. (2004) Santa Croce. Interpretazione attraverso le indagini metriche e documentarie. In: Rocchi Coopmans de Yoldi, G. (ed.), *S.Maria del Fiore e le chiese fiorentine del Duecento e del Trecento nella città delle fabbriche arnolfiane*, 243–262.

Carli, E. (1979) *Il duomo di Siena*. Genova, Sagep.

Casini, C. (1983) *I restauri del Duomo di Pisa 1959–1981*, degree thesis in History of Architecture, supervisor Prof. Giacinto Nudi, University of Pisa, Faculty of Literature and Philosophy, academic year 1982–1983, tome II, part II, 955- 956.

Causarano, M.A. (2009) La cattedrale e la città: il cantiere del duomo di Siena. Risultati delle indagini archeologiche. *Arqueologia de la Arquitectura*, 6, enero-diciembre 2009, 207.

Causarano, M.A. (2017) *La cattedrale e la città. Il cantiere del duomo di Siena tra XI e XIV secolo*. Sesto Fiorentino, All'Insegna del Giglio, 100–102.

Causarano, M.A., Francovich, R. & Valenti M. (2003) L'intervento archeologico sotto il duomo di Siena: dati e ipotesi preliminari. In: Guerrini, R. (ed.), *Sotto il duomo di Siena. Scoperte archeologiche, architettoniche e figurative*. Milano, Silvana Editoriale, 153–168.

Cavallucci, C.J. (1881) *S. Maria del Fiore. Storia documentata dall'origine fino ai nostri giorni*. Firenze, Cirri.

Cervini, F. & De Marchi, A. (eds.) (2016) *Duomo di Siena 1300. Affreschi,policromie, apparati: un network di immagini*. Roma, Carocci.

Consiglio nazionale delle ricerche: Gruppo catalogo dei terremoti (1985) *Catalogo dei terremoti italiani dall'anno 1000 al 198*. Roma, CNR.

Cresti, C. & Zangheri, L. (1978) *Architetti e ingegneri nella Toscana dell'Ottocento*. Firenze, Uniedit.

Del Moro, L. (1895) *Relazione sui danni arrecati ai monumenti insigni dal terremoto del 18 maggio 1895*. Firenze, Carnesecchi.

Facoltà di Architettura: Università degli Studi di Firenze (1970) *Il Duomo di Pisa: rilievo a cura dell'Istituto di restauro dei monumenti*. Pisa, Nistri Lischi.

Galli, L. (2006) Danni e restauri ottocenteschi alla "copertura" del Duomo di Siena. In: Lorenzoni, M. (ed.), *Forte fortuna, religiosità e arte nella cultura senese dalle origini all'umanesimo di Pio II ai restauri del XIX secolo; leggere l'arte della chiesa*, 2. Siena, OPA, 347–405.

Garzonio, C.A. et al. (2010) Analisi dei campioni prelevati dalla cupola del Duomo di Siena e dalle strutture limitrofe. In: Giorgi & Matracchi, Cupole medievali, 117–118.

Gennari, M. (2005) *L'orribil scossa della vigilia di pentecoste: Siena e il terremoto del 1798*. Monteriggioni, Il Leccio.

Giorgi, A. & Moscadelli, S. (2003) "Ut homines et persone possint comode ingredi". Direttrici viarie e accessi orientali del duomo di Siena nella documentazione dei secoli XII e XIII. In: Guerrini, R. (ed.), *Sotto il duomo di Siena. Scoperte archeologiche, architettoniche e figurative*. Milano, Silvana Editoriale, 90–105.

Giorgi, A. & Moscadelli, S. (2005) *Costruire una cattedrale. L'Opera di Santa Maria di Siena tra XII e XIV secolo*. München, Deutscher Kunstverlag.

Giorgi, A. & Moscadelli, S. (2019) Fonti documentarie. In: Bartalini, R., *Il Duomo Nuovo di Siena. La fabbrica, le sculture, i maestri, le dinamiche di cantiere*. Milano, Silvana, 178–186.

Giorgi, L. (1988a) La composizione della facciata in elevato. In: Rocchi, G. et al., *S. Maria del Fiore: Rilievi, documenti, indagini strumentali. Interpretazione. Il corpo basilicale*, 40–43.

Giorgi, L. (1988b) Le vicende della facciata dalle origini all'Ottocento. In: Rocchi, G. et al., *S. Maria del Fiore: Rilievi, documenti, indagini strumentali. Interpretazione. Il corpo basilicale*, 33–39.

Giorgi, L. (1988c) Lo sviluppo del corpo basilicale. In: Rocchi, G. et al., *S. Maria del Fiore: Rilievi, documenti, indagini strumentali. Interpretazione. Il corpo basilicale*, 44–50.

Giorgi, L. (2017) Aspetti costruttivi del Campanile di Giotto. In: Gurrieri, F. (ed.), *Il Campanile di Giotto. Atti del ciclo di Conferenze, Firenze, maggio-giugno 2015*. Firenze, Mandragora, 15–32.

Giorgi, L. & Marino, L. (1987) Accertamenti sulla originaria stesura delle prime campate di Santa Maria del Fiore. In: Spagnesi, G. (ed.), *Esperienze di Storia dell'Architettura e di Restauro, Atti del*

XXI Congresso di Storia dell'Architettura (Roma, 12–14 ottobre 1983). Roma, Istituto della Enciclopedia Italiana, 63–68.

Giorgi, L. & Marino, L. (1988) I caratteri del sistema voltato. In: Rocchi, G. et al., *S. Maria del Fiore: Rilievi, documenti, indagini strumentali. Interpretazione. Il corpo basilicale*, 51–60.

Giorgi, L. & Matracchi, P. (2006a) I reperti ipogei delle primitive Santa Maria Novella, Santa Croce, Santa Maria del Fiore, Sant'Egidio. In: Rocchi Coopmans de Yoldi, G. (ed.), *S. Maria del Fiore. Teorie e storie dell'archeologia e del restauro nella città delle fabbriche arnolfiane*, 97–112.

Giorgi, L. & Matracchi, P. (2006b) Santa Maria del Fiore, Facciata, corpo basilicale, cupola. In: Rocchi Coopmans de Yoldi, G. (ed.), *S. Maria del Fiore. Teorie e storie dell'archeologia e del restauro nella città delle fabbriche arnolfiane*, 277–324.

Giorgi, L. & Matracchi, P. (2010) *Cupole medievali. Il Duomo di Siena. I. Diagnostica architettonica per il cantiere di restauro / Medieval Domes, the Cathedral of Siena, Architectural diagnostics for the restoration works, I*. Firenze, Alinea.

Giorgi, L. & Matracchi, P. (2011). La chiesa di Santa Croce e i precedenti insediamenti francescani. Architettura e resti archeologici. In: De Marchi, A. & Piraz, G. (eds.), *Santa Croce. Oltre le apparenze*. Pistoia, Gli Ori, 12–31.

Giorgi, L. & Matracchi, P. (2012) Stability of the structure of the basilican body of Santa Maria del Fiore in Florence during construction of the naves. In: Jasieńko, J. (ed.), *Structural Analysis of Historical Constructions, Proceedings of the International Conference on Structural Analysis of Historical Construction, SACH 2012, 15-17 October, Wrocław, 2*. Wroclaw, DWE, 1127–1135.

Giorgi, L. & Matracchi, P. (2017) Le murature a cassone alla base della cupola del Battistero e altri aspetti costruttivi. In: Gurrieri, F. (ed.), *Il Battistero di San Giovanni. Conoscenza, diagnostica, conservazione*. Firenze, Mandragora, 193–207.

Giorgi, L. & Matracchi, P. (2019) *Le torri di San Gimignano. Architettura, città, restauro/The towers of San Gimignano. Architecture, town, restoration*. Firenze, didapress.

Giorgi, L., Matracchi, P. & Rocchi Coopmans de Yoldi, G. (2009). Architettura e pittura nella cripta ionica del duomo di Siena. *Commentari d'arte*, XIV, n. 41, 5–29.

Gli Anni della Cupola. Archivio digitale delle fonti dell'Opera di Santa Maria del Fiore, 1417–1436 (2015) Haines, M. (ed.). Firenze, Opera di Santa Maria del Fiore, http://archivio.operaduomo.fi.it/cupola/.

Guasti, C. (1887) *Santa Maria del Fiore: la costruzione della chiesa e del campanile secondo i documenti tratti dall'archivio dell'Opera Secolare e da quello di Stato*. Firenze, Tip. M. Ricci.

Haas, W. & von Winterfeld, D. (1999a) *Der Dom S. Maria Assunta: Architektur 3.1.2: Bildband*. Munchen, Bruckmann.

Haas, W. & von Winterfeld, D. (1999b) *Der Dom S. Maria Assunta: Architektur 3.1.3: Planband*. Munchen, Bruckmann.

Haas, W. & von Winterfeld, D. (2006a) *Der Dom S. Maria Assunta : Architektur 3.1.1.1: Textband*. Munchen, Bruckmann.

Haas, W. & von Winterfeld, D. (2006b) *Der Dom S. Maria Assunta : Architektur 3.1.1.2: Textband*. Munchen, Bruckmann.

Haines, M. (2013) Myth and management in the construction of Brunelleschi's Cupola. *I Tatti Studies*, 14–15/2011–2012, 47–101.

Leoncini, A. (2005) Considerazioni sulla confessione del duomo di Siena. In: Lorenzoni, M. (ed.) *Forte fortuna, religiosità e arte nella cultura senese dalle origini all'umanesimo di Pio II ai restauri del XIX secolo; leggere l'arte della chiesa, 2*. Siena, OPA, 459–482.

Levi, D. (2015) Igino Benvenuto Supino e il restauro ottocentesco del Duomo di Pisa. In: Angiolini, F. & Baldassarri, M. (eds) *I Supino, una dinastia di ebrei pisani fra mercatura, arte, politica e diritto (secoli XVI-XX)*. Pisa, Pacini, 125–127.

Marini, M. (1981) Ricostruzione della Controcupola del Duomo di Siena. 1890–1892. In: Buscioni, M.C. (ed.), *Giuseppe Partini: Architetto del Purismo senese*. Firenze, Electa, 186.

Marino, L. (1988) Interpretazione delle murature ipogee, dei reperti archeologici, degli scavi, delle analisi di laboratorio. In: Rocchi, G. et al., *S. Maria del Fiore: Rilievi, documenti, indagini strumentali. Interpretazione. Il corpo basilicale*, 15–29.

Marino, L., Dinelli, O., Guerra, M., Nenci, C. & Orlando, F. (1996) L'ipogeo del Battistero di Firenze. In: Rocchi Coopmans de Yoldi, G. (ed.), *S. Maria del Fiore. Piazza, Battistero, Campanile*, 69–72.

Masturzo, N. (2006) Santa Reparata e Santa Maria del Fiore. Studio delle fondazioni in facciata. In: Rocchi Coopmans de Yoldi, G. (ed.), *S. Maria del Fiore. Teorie e storie dell'archeologia e del restauro nella città delle fabbriche arnolfiane*, 209–222.

Matracchi, P. (2006). La diffusione delle Chiese a sala a Perugia: L'edificazione della Cattedrale di San Lorenzo coeva al San Domenico. In: Rocchi Coopmans De Yoldi, G. & Ser-Giacomi, G. (eds.), *La Basilica di San Domenico a Perugia*. Perugia, Quattroemme, 120–150.

Matracchi, P. (2013). Roberto Di Stefano e la cupola di San Pietro: storiografia e restauro. In: Aveta, A. & Di Stefano, M.(eds.), *Roberto di Stefano, filosofia della conservazione e prassi del restauro*. Napoli, Arte Tipografica Editrice, 185–190.

Matracchi, P. & Giorgi, L. (2012) New insights on the structural weakness of the Duomo Nuovo in Siena. In: Jasieńko, J. (ed.), *Structural Analysis of Historical Constructions, Proceedings of the International Conference on Structural Analysis of Historical Construction, SACH 2012, 15-17 October, Wrocław, 1*. Wrocław, DWE, 323-324.

Matracchi, P., Radicioni, F., Stoppini, A. & Tosi, G. (2019) The Cathedral of S. Lorenzo in Perugia and the hypogeal spaces. Geomatic Techniques for Spatial Investigations Aimed at the Knowledge and Interpretation of the Origin of the Transept. *The International Archives of the Photogrammetry, Remote Sensing and Spatial Information Sciences*, XLII-2/W11, 849-856.

Miceli, E. & Papi, R. (2004) Il comportamento statico delle navate della Cattedrale di Santa Maria del Fiore. In: Rocchi Coopmans de Yoldi, G. (ed.), *S. Maria del Fiore e le chiese fiorentine del Duecento e del Trecento nella città delle fabbriche arnolfiane*, 311–320.

Milone, A. (2008) Il Duomo di Pisa. Arte e città nel XII secolo. In: *Storie di artisti, storie di libri: l'editore che inseguiva la bellezza: scritti in onore di Franco Cosimo Panini*. Roma, Donzelli, 73–74, 78.

Morelli, A. (2006) Indagini diagnostiche di tipo radar applicate allo studio di sezioni murarie e alla ricerca di strutture archeologiche nascoste e ipogee. In: Rocchi Coopmans de Yoldi, G. (ed.), *S. Maria del Fiore. Teorie e storie dell'archeologia e del restauro nella città delle fabbriche arnolfiane*, 223–228.

Morozzi, G. (1987) *Santa Reparata. L'antica cattedrale fiorentina*. Firenze, Bonechi.

Morozzi, G., Toker, F. & Herrmann, J. (1974) *Santa Reparata. L'antica cattedrale fiorentina. I risultati dello scavo condotto dal 1965 al 1974*. Firenze, Bonechi.

Nenci, C. (1996) Gli scavi nella piazza e all'interno del Battistero e del Duomo. In: Rocchi Coopmans de Yoldi, G. (ed.), *S. Maria del Fiore. Piazza, Battistero, Campanile*, 15–26.

Pedretti, C. (1988) *Leonardo architetto*. Milano, Electa.

Peroni, A. (1995) Architettura e decorazione. In: Peroni, A. (ed.), *Il Duomo di Pisa*, 13–147.

Peroni, A. (ed.) (1995) *Il Duomo di Pisa*. Modena, Panini.

Pietramellara, C. (1980) *Il Duomo di Siena: Evoluzione della forma dalle origini alla fine del Trecento*. Firenze, EDAM.

Pomarici, F. (2004) *La prima Facciata di Santa Maria del Fiore. Storia e interpretazione*. Roma, Viella, 60.

Quast, M. (2007) La facciata occidentale del Duomo Vecchio: l'architettura. In: Lorenzoni, M. (ed.) *La facciata del Duomo di Siena: iconografia, stile, indagini storiche e scientifiche*. [Siena], Banca Monte dei Paschi di Siena, 97–129.

Riccetti, L. (2001) Il vescovo Francesco Monaldeschi e l'avvio del cantiere di Santa Maria del Fiore (1295–1301). In: Verdon, T. & Innocenti, A. (eds.), *La cattedrale e la città. Saggi sul duomo di Firenze, Atti del Convegno Internazionale di Studi, 16–21 giugno 1997, Firenze*, I. Firenze, Edifir, 219–225.

Roca, P., Lourenço, P.B. & Gaetani A. (2019) *Historic Construction and Conservation. Materials, Systems and Damage*. New York & London: Routledge, an imprint of the Taylor & Francis Group.

Rocchi Coopmans De Yoldi, G. (1996a) Interpretazione del campanile. In: Rocchi Coopmans de Yoldi, G. (ed.), *S. Maria del Fiore. Rilievi, documenti, indagini strumentali, interpretazioni. Piazza, Battistero, Campanile*, 129–144.

Rocchi Coopmans De Yoldi, G. (1996b) La fondazione di Santa Maria del Fiore e la facciata trecentesca. In: Rocchi Coopmans de Yoldi, G. (ed.), *S. Maria del Fiore. Rilievi, documenti, indagini strumentali, interpretazioni. Piazza, Battistero, Campanile*, 155–157.

Rocchi Coopmans De Yoldi, G. (2004) Il complesso di Santa Maria del Fiore. In: Rocchi Coopmans de Yoldi, G. (ed.), *S. Maria del Fiore e le chiese fiorentine del Duecento e del Trecento nella città delle fabbriche arnolfiane*, 93–112.

Rocchi Coopmans De Yoldi, G. (2006a) Il cantiere del complesso di Santa Maria del Fiore dall'epoca arnolfiano-giottesca a quella brunelleschiana. In: Rocchi Coopmans de Yoldi, G. (ed.), *S. Maria del Fiore. Teorie e storie dell'archeologia e del restauro nella città delle fabbriche arnolfiane*, 243–276.

Rocchi Coopmans De Yoldi, G. (2006b) Teorie e storie dell'architettura e del restauro. In: Rocchi Coopmans de Yoldi, G. (ed.), *S. Maria del Fiore. Teorie e storie dell'archeologia e del restauro nella città delle fabbriche arnolfiane*, 361–414.

Rocchi Coopmans De Yoldi, G. (2006c). Dinamica delle chiese degli Ordini mendicanti e predicatori. In: Rocchi Coopmans De Yoldi, G. & Ser-Giacomi, G. (eds.), *La Basilica di San Domenico a Perugia*. Perugia, Quattroemme, 151–176.

Rocchi Coopmans de Yoldi, G. (ed.) (1996) *S. Maria del Fiore: Rilievi, documenti, indagini strumentali. Interpretazioni. Piazza, Battistero, Campanile*. Firenze, Il Torchio.

Rocchi Coopmans de Yoldi, G. (ed.) (2004) *S. Maria del Fiore e le chiese fiorentine del Duecento e del Trecento nella città delle fabbriche arnolfiane*. Firenze, Alinea.

Rocchi Coopmans de Yoldi, G. (ed.) (2006) *S. Maria del Fiore. Teorie e storie dell'archeologia e del restauro nella città delle fabbriche arnolfiane*. Firenze, Alinea.

Rocchi, G. (1988) Le fasi della fabbrica dal periodo arnolfiano a quello talentiano. In: Rocchi, G. et al., *S. Maria del Fiore: Rilievi, documenti, indagini strumentali. Interpretazione. Il corpo basilicale*, 63–72.

Rocchi, G., Bebber, A., Franchi, R., Giorgi, L. & Marino, L. (1988) *S. Maria del Fiore: Rilievi, documenti, indagini strumentali. Interpretazione. Il corpo basilicale*. Milano, Hoepli.

Ronzani, M. (1996) Dall'edificatio ecclesiae all'«Opera di S. Maria»: nascita e primi sviluppi di un'istituzione nella Pisa dei secoli XI e XII. In: Haines, M. & Riccetti, L. (eds.), *Opera. Carattere e ruolo delle fabbriche cittadine fino all'inizio dell'Età Moderna, atti della Tavola Rotonda, Villa I Tatti, Firenze (3 aprile 1991)*. Firenze, L. S. Olschki, 35.

Rossi, C., Baldi, A.M. & Ricci, F. (2003) Appendice 3. Indagini Geologiche: esplorazione e caratterizzazione del sottosuolo. In: Guerrini, R. (ed.) *Sotto il duomo di Siena. Scoperte archeologiche, architettoniche e figurative*. Milano, Silvana Editoriale, 216–221.

Saalman, H. (1980) *Filippo Brunelleschi. The Cupola of Santa Maria del Fiore*. London, Zwemmer.

Sanpaolesi, P. (1959) Il restauro delle strutture della cupola della Cattedrale di Pisa. *Bollettino d'Arte*, III, 207.

Sanpaolesi, P. (1975) *Il duomo di Pisa e l'architettura romanica toscana delle origini*. Pisa, Nistri-Lischi.

Scalia, G. (1982) Tre iscrizioni e una facciata. Ancora sulla cattedrale di Pisa. *Studi medievali*, s. III, XXIII, 817–859.

Seidel, M. (2003) Tradizione e innovazione. Note sulle scoperte architettoniche nel duomo di Siena. In: Guerrini, R. (ed.), *Sotto il duomo di Siena. Scoperte archeologiche, architettoniche e figurative*. Milano, Silvana Editoriale, 35–84.

Sembranti, S. (1997) La fabbrica del duomo di Siena. Restauri consolidativi e comportamento strutturale. *Bullettino Senese di Storia Patria*, 102, 1995, 63–158.

Tanfani Centofanti, L. (1897) *Notizie di artisti tratte dai documenti Pisani*. Pisa, Spoerri (reprint 1972, Bologna, Forni).

Tigler, G. (2007) Siena 1284–1297. Giovanni Pisano e le sculture della parte bassa della facciata. In: Lorenzoni, M. (ed.), *La facciata del Duomo di Siena: iconografia, stile, indagini storiche e scientifiche*. [Siena], Banca Monte dei Paschi di Siena, 131–145.

Tigler, G. (2009) Il pulpito di Nicola Pisano. In: Lorenzoni, M. (ed.), *Le sculture del duomo di Siena*. Cinisello Balsamo, Silvana, 122–131.

Toesca, P. (1965) *Il Medioevo, II, Il Trecento*. Torino, UTET.

Tosti, G. (2006) Studio diagnostico del Duomo di Siena. Relazione tecnica preliminare. Commissione tecnico scientifica incaricata dall'Opera Metropolitana del Duomo di Siena, Report.

Tragbar, K. (1997) Il campanile del duomo di Siena e le torri gentilizie della città. *Bullettino senese di storia patria*, 102, 1995, 159–186.

Tragbar, K. (2018) Siena 1357: the failure of a great plan. In: Wouters, I., Van de Voorde, S., Bertels, I. et al. (eds.) *Building knowledge, constructing histories : proceedings of the Sixt International Congress on Construction History (6ICCH), Brussels, Belgium, 9-13 july 2018*, I. London, CRC, 43-50.

Villani, G. (1991) Nuova Cronica. Parma, Fondazione Pietro Bembo/Ugo Guanda, libro X, Cap. 193.

Index of subjects

Index of names and places

Note: *Italic* page numbers refer to figures.

Milton Keynes UK
Ingram Content Group UK Ltd.
UKHW052028141024
449569UK00017B/739